SMART VILLAGES IN TBEYOND

EMERALD STUDIES IN POLITICS AND TECHNOLOGY

Series Editors: Anna Visvizi and Miltiadis D. Lytras

This series focuses broadly on the intersection of politics and technology. Its objective is to identify and explore the critical junctions where politics and information and communication technology (ICT) intersect to showcase the opportunities, raise awareness, and preempt impending risks for our societies.

The series has a broad scope and addresses a variety of topics, including but not limited to the following: cyber-intelligence; government analytics; user-generated data and its impact on human society; technology in healthcare and public services; quantitative measures in political discourse; public engagement with politics through technology – for example, blogs, social media; freedom of the internet; text mining; e-participation in politics and digital diplomacy; international trade on the ICT market; information security risks; political communication in online social networks; big data; e-government and e-democracy; digital activism; ICT in developing nations; digital media; smart cities; disruptive effects of technology in politics; internet governance; citizen journalism; the politics of migration and ICT; and the European Union and ICT.

We are actively seeking proposals for this exciting new series – please contact the editors if you are interested in publishing in this series.

Interested in publishing in this series? Please contact Anna Visvizi and Miltiadis D. Lytras, avisvizi@acg.edu and mlytras@acg.edu.

SMART VILLAGES IN THE EU AND BEYOND

EDITED BY

ANNA VISVIZI
Deree College – The American College of Greece

MILTIADIS D. LYTRAS
Deree College – The American College of Greece

GYÖRGY MUDRI
European Parliament

United Kingdom – North America – Japan – India – Malaysia – China

Emerald Publishing Limited
Howard House, Wagon Lane, Bingley BD16 1WA, UK

First edition 2019

Copyright © 2019 Editorial matter and selection © Anna Visvizi, Miltiadis D. Lytras and György Mudri, published under exclusive licence. Individual chapters © the respective Authors.

Reprints and permissions service
Contact: permissions@emeraldinsight.com

No part of this book may be reproduced, stored in a retrieval system, transmitted in any form or by any means electronic, mechanical, photocopying, recording or otherwise without either the prior written permission of the publisher or a licence permitting restricted copying issued in the UK by The Copyright Licensing Agency and in the USA by The Copyright Clearance Center. Any opinions expressed in the chapters are those of the authors. Whilst Emerald makes every effort to ensure the quality and accuracy of its content, Emerald makes no representation implied or otherwise, as to the chapters' suitability and application and disclaims any warranties, express or implied, to their use.

British Library Cataloguing in Publication Data
A catalogue record for this book is available from the British Library

ISBN: 978-1-78769-846-8 (Print)
ISBN: 978-1-78769-845-1 (Online)
ISBN: 978-1-78769-847-5 (Epub)
ISBN: 978-1-78769-848-2 (Paperback)

Printed and bound by CPI Group (UK) Ltd, Croydon, CR0 4YY

ISOQAR certified Management System, awarded to Emerald for adherence to Environmental standard ISO 14001:2004.

Certificate Number 1985
ISO 14001

INVESTOR IN PEOPLE

Contents

List of Figures — vii

List of Tables — ix

About the Authors — xi

Editors' Preface — xv

Foreword — xvii

Chapter 1 Smart Villages: Relevance, Approaches, Policymaking Implications
Anna Visvizi, Miltiadis D. Lytras and György Mudri — 1

Chapter 2 Integrated Approach to Sustainable EU Smart Villages Policies
Christiane Kirketerp de Viron and György Mudri — 13

Chapter 3 Smart Villages Revisited: Conceptual Background and New Challenges at the Local Level
Oskar Wolski and Marcin Wójcik — 29

Chapter 4 Toward a New Sustainable Development Model for Smart Villages
Raquel Pérez-delHoyo and Higinio Mora — 49

Chapter 5 The Role of LEADER in Smart Villages: An Opportunity to Reconnect with Rural Communities
Enrique Nieto and Pedro Brosei — 63

Chapter 6 Precision Agriculture and the Smart Village Concept
Daniel Azevedo — 83

Chapter 7 Energy Diversification and Self-sustainable Smart Villages
James K. R. Watson   *99*

Chapter 8 The Role of Smart and Medium-sized Enterprises in the Smart Villages Concept
Xénia Szanyi-Gyenes   *111*

Chapter 9 Smart Villages in Slovenia: Examples of Good Pilot Practices
Veronika Zavratnik, Andrej Kos and Emilija Stojmenova Duh   *125*

Chapter 10 Smart Village Projects in Korea: Rural Tourism, 6th Industrialization, and Smart Farming
Jonghoon Park and Seongwoo Lee   *139*

Chapter 11 Smart Villages and the GCC Countries: Policies, Strategies, and Implications
Tayeb Brahimi and Benaouda Bensaid   *155*

Chapter 12 Smart Villages: Mapping the Emerging Field and Setting the Course of Action
Miltiadis D. Lytras, Anna Visvizi and György Mudri   *173*

Index   *177*

List of Figures

Chapter 1

Figure 1.	Smart Village: The Three Pillars of the Comprehensive Approach to Smart Village.	4
Figure 2.	Smart Villages: From Needs to Targeted and Effective Policymaking.	5

Chapter 4

Figure 1.	A Methodology for the Design of a Smart Strategy. . . .	54
Figure 2.	The 'Three Magnets' Theory – Garden City Model.	56
Figure 3.	Benefits of Living in a Smart Village..	57

Chapter 5

Figure 1.	LEADER Principles..	68
Figure 2.	Potential Actions in Smart Village Plans..	71
Figure 3.	Mapping Rural Innovation and Policy Tools.	73

Chapter 10

Figure 1.	Basic Concepts of 6th Industrialization.	149

Chapter 11

Figure 1.	GCC Population Distribution and Estimation (2018–2030)..	160
Figure 2.	GCC Population, Urban and Rural, in 2018..	161
Figure 3.	GCC Population, Urban and Rural, in 2018, 2025, and 2030	161

List of Tables

Chapter 1

Table 1.   Smart Villages: Typology of Challenges and the Corresponding Urgency of Action................. 2

Chapter 5

Table 1.   Distinctive Characteristics between Smart Villages and LEADER........................ 69

Chapter 10

Table 1.   Agricultural and Rural Investment Plans in Korea (1991 to 2017)...................... 144

Table 2.   Rural Tourism Programs in Korea in the Post-productivist Era..................... 146

Chapter 11

Table 1.   GCC Population, Urban and Rural Areas.......... 160

About the Authors

Daniel Azevedo, Director at Copa and Cogeca, graduated in Biophysics Engineering – Environmental Management and Planning from Universidade deÉvora, Portugal. Daniel Azevedo is Director in the Commodities and Trade team of Copa and Cogeca, an agricultural lobby representing almost 70 national farm organizations and cooperatives in Europe. He is a natural resources engineer graduating from the University of Évora (Portugal) and currently is the Coordinator of Copa and Cogeca Task Force on agricultural technology. He was previously working for DG Agriculture and Rural Development after specializing in agro-environmental measures in SLU University (Sweden). Mr Azevedo comes from a family producing wine and olive oil in Vila Real (Douro region) in Portugal.

Benaouda Bensaid, PhD, is Associated Faculty at College of Art and Science, Effat University, Jeddah, Saudi Arabia. Dr Bensaid earned his PhD from the Institute of Islamic Studies, McGill University, Canada.

Tayeb Brahimi, PhD, is Assistant Professor at College of Engineering, Effat University, Jeddah, Kingdom of Saudi Arabia (KSA). He received his PhD (1992) from Ecole Polytechnique de Montreal, Canada. His current research interests relate to renewable energy, sustainability, green engineering, engineering education, quality assurance, and integrating Islamic innovative heritage into the higher education curricula.

Pedro Brosei is Senior Local and Rural Development Expert, Thematic & Territorial Coordinator at the Fisheries Areas Network (FARNET) Support Unit, and Advisor (volunteer) to the Portuguese Presidency of the European LEADER Association for Rural Development (ELARD) during 2018–2019. Pedro was also Vice-president of ELARD (2016–2017). Previously, he was Horizontal Coordinator for LEADER/CLLD in the European Commission DG AGRI from 2008 to 2014. During this time, he was responsible for the analysis and development of the LEADER/CLLD approach within DG AGRI. Prior to that, he worked as a Coordinator in the European LEADER + Observatory (2004–2008) and the German LEADER+ Network Unit (2002–2004).

Christiane Kirketerp de Viron is Civil Servant at the European Commission. Christiane is a Political Scientist. She joined the European Commission in 2006 and currently serves as Member of Cabinet to the Commissioner for Research Science and Innovation. Prior to this, she was responsible for the conception of rural development policy and for the development of the smart villages initiative within the Directorate General for Agriculture and Rural Development.

Andrej Kos, PhD, is Professor at the Faculty of Electrical Engineering, UL, and Head of Laboratory for telecommunications. His research work focuses on IoT, digitalization, broadband networks (in rural areas) and the use of distributed ledger technologies in industry. He is Head of Innovation Commission at the University of Ljubljana.

Lee, Seongwoo is Professor at Seoul National University in Korea since 1998. He received his PhD degree in Planning from the University of Southern California in USA. His research interests are rural development strategy, housing, policy evaluation, and spatial econometrics models.

Miltiadis D. Lytras, PhD, is Research Professor at Deree College – The American College of Greece, and Visiting Researcher at Effat University. Researcher, Editor, Lecturer, and Consultant, Dr Lytras' expertise covers issues pertinent to the broad field defined by cognitive computing, information systems, technology-enabled innovation, social networks, computers in human behavior, and knowledge management. In his work, Dr Lytras focuses on bringing together advances in ICT and knowledge management to advance socioeconomic sustainability and citizens' well-being.

Higinio Mora received his PhD degree in Computer Science from the University of Alicante (Spain) in 2003. His areas of research interest include computer modeling, embedded systems, internet of things, and cloud computing paradigm. His work has been published in international journals and conferences, with more than 100 published papers.

György Mudri is MSc Agronomist, majoring in both Genetics and Biotechnology from Szent István University of Hungary and Stuttgart-Hohenheim; he has a NLD HBO Engineers Degree in International Rural Innovation and Development from the Netherlands. He is Rural Development Expert, working as Advisor and Accredited Parliamentary Assistant (APA) in the European Parliament. Prior to that, he was a Candidate MEP. He has also worked as Policy Officer at the European Commission (DG AGRI), and was Personal Secretary and later Advisor to the Candidate Prime Minister in Hungary and Policy Officer at the Ministry of Agriculture and Rural Development in Hungary.

Enrique Nieto is Senior Expert on territorial development policies in the European Association for Information on Local Development (AEIDL). As AEIDL expert, Enrique is engaged in the European Network for Rural Development (ENRD) since 2015 as Policy Analyst. Previously, he was Consultant for the Food and Agriculture Organisation of United Nations (FAO; 2012), Evaluation Officer in the Evaluation Helpdesk of the ENRD (2013–2014), and Policy Officer in the Fisheries Areas Network (FARNET; 2014–2015).

Park, Jonghoon received his PhD degree in Economics from Seoul National University in February 2019. He is Lecturer in SungKyul University in Korea.

His research interests are in the fields of regional development and planning focusing on rural policy, housing, aging and welfare.

Raquel Pérez-delHoyo completed her PhD from the University of Alicante. Dr Pérez-delHoyo is Architect and Specialist in urban planning, with a PhD in Architecture, City, Civil Works, and Their Construction. She is Lecturer at Urban Design and Regional Planning Unit, University of Alicante. She develops research on urban planning, smart cities, and inclusive cities. Her main area of interest is humanization of smart cities, including the development of models focused on people to improve the design and planning of smart cities.

Emilija Stojmenova Duh, PhD, is Assistant Professor at the Faculty of Electrical Engineering, UL. Her research work focuses on user centred design, design thinking, open innovation and digitalisation for rural development. She is Coordinator of FabLab Network Slovenia and Director of Digital Innovation Hub Slovenia.

Xenia Szanyi-Gyenes is PhD Candidate at the University of Corvinus, Budapest. Her thesis is about 'The Prospects and Opportunities of Small Companies to Enter the International Market.' She has gained relevant work experience as Investment Adviser at a venture capital fund by working with small- and medium-sized enterprises.

Anna Visvizi, PhD, is Associate Professor at Deree College – The American College of Greece, and Visiting Researcher at Effat University. Researcher, Editor, Policy Advisor, and Lecturer, Dr Visvizi's expertise covers issues pertinent to the intersection of politics, economics, and ICT, including multilateralism and international organizations (especially the European Union, NATO, the OECD, and the WTO), smart cities and smart villages, and migration. In her work, Dr Visvizi places emphasis on engaging academia, the think-tank sector and decision-makers in dialogue to ensure well-founded and evidence-driven policymaking.

James K.R. Watson is Secretary General of Eurogas, the European gas industry association, since January 2019. Previously he was Chief Executive Officer of Solar Power Europe, the European solar industry association. He has worked for the Commonwealth Secretariat on a European Commission project on trade and sustainable development in Ethiopia and was Lecturer in Environmental Law at the University of Manchester. He holds a PhD in International Trade and Environmental Law from the University of Leeds, and is Visiting Professor at the Vrije Universiteit Brussel.

Marcin Wójcik, PhD, is Professor at the University of Lodz, Faculty of Geographical Sciences, Department of Regional and Social Geography. He is the Author of publications on rural development, cultural landscape, local development, and socio-spatial diversities, and Manager of national and international projects. He is also Chairman of the Commission of Rural Areas of the Polish Geographical Society, Member of the Lodz Scientific Society and the Task

Force for Rural Areas and Landscape of the Committee for Spatial Economy and Regional Planning (Polish Academy of Sciences).

Oskar Wolski is PhD Candidate at the University of Lodz, Faculty of Geographical Sciences, Department of Regional and Social Geography. He is interested in rural development, local and regional development, and the selected issues of rural geography. He is Member of the Commission of Rural Areas of the Polish Geographical Society, and Expert of the European Network for Rural Development (ENRD) on Smart and Competitive Rural Areas (Thematic Group on Smart Villages) and of the Agricultural European Innovation Partnership (EIP-AGRI) on digitalization of rural areas.

Veronika Zavratnik is PhD Candidate at the University of Ljubljana, Faculty of Arts, Department of Ethnology and Cultural Anthropology. She is Researcher at the Laboratory for telecommunication, Faculty of Electrical Engineering, UL. Her research work focuses on material culture, rural development, smart communities, cultural heritage, and digital anthropology.

Editors' Preface

Connecting the slopes of a valley hidden in the mountains of Arcadia (Peloponnese, Greece), the arch bridges in Fouskari stretching over the waters brought by three springs, are suggestive of how life might have looked like just a few decades back. The stone-curved aqueducts and tiny water basins remind us of the lifestyle and hardship the village inhabitants endured. But the view of the green slopes of the valley also seem to be reviving the laughter of girls and women meeting here in the past to do laundry or take a bath. *Et in Arcadia Ego* is a potent title of a seventeenth century painting by Nicolas Poussin. The idyllic representation of shepherds and the mystery that the painting conveys remind us of Virgil's *Arcadia* and the archetypal pastoral milieu. Explored and described by Pausanias (110–180 AD), a land filled with treasures still waiting to be explored, Arcadia is a land bursting with myths of Gods, nymphs, and good charms. Located in direct vicinity of Ancient Olympia, it is a land poignant with thousand years of history, heroism, and courage curved in the walls of abandoned castles, and told over and over again by trees and rivers. But Arcadia is depopulating rapidly today. This book stems from our concern about Arcadia and its inhabitants; it derives from our commitment to revitalize the area and from our hope that it is feasible.

This book would not be possible without the Publisher who cordially embraced the book idea, the reviewers, and the contributing authors. Special 'thank you' is extended to the Editorial Assistant, Ms Anna Scaife, and the entire Emerald Publishing team.

We dedicate this book to Arcadia and the memories it brings,
The Editors: Anna Visvizi, Miltiadis D. Lytras, and György Mudri

Foreword

Smart Villages Approach for a Brighter Future of Rural Communities in the European Union

The European Union (EU) is an exemplary smart model, constantly investing in innovation and development. Along with this smart model and beyond the pure economic advantages, the quality of life of people has always been the key force behind the EU policymaking. However, the impact of these efforts has been different in urban and rural areas. Therefore, today, there is a clear need for a new integrated, innovative approach to rural areas in the EU. This approach can be best termed as 'smart villages' approach.

The expansion of ICT-enhanced applications and services enable societies to improve their opportunities and improve their attractiveness and the quality of life not only in urban but in rural areas as well. We see many twenty-first century innovations in our constituencies; innovations, that are usually a little bit more and a little bit quicker in urban areas. People dealing with rural areas sense it for a while, even before the concrete actions appeared in the form of the smart villages concept.

Of course, one may wonder: what exactly does 'smart' mean for us as politicians in the smart villages concept? Is it life, water, energy, community, or food? Is it the technology, the ways and means, or the status? What do villages, or rural areas in the concept actually stand for?

Our first answer is that there are different smart elements, which definitely share some common layers. Their meaning, however, may differ in different parts of our globe. But being smart definitely is about intelligent applications, the various interactions of the existing and new technologies, and also the efficient use of big data analysis adjustments. The concept of smart villages does not propose a one-size-fits-all solution. It is territorially sensitive, based on the needs and potential of a given territory, and strategy-led, supported by new or existing territorial strategies. In addition, when talking about European villages, and rural areas, we do not solely mean the 22 million EU farmers, or people working directly in the agricultural sector. More than half of the EU's land area is within regions classified as predominantly rural. More than 112 million people inhabit these areas. We are happy to welcome and commend many EU objectives under different funds and policies, such as the Common Agricultural Policy, related to innovation, digitalization, transformation, and modern rural life in the EU.

At the same time, we can extend the approach to become global, as rural areas face specific challenges that need specific solutions everywhere. Aging populations, lack of services (medical, postal, health, transport, and energy), and limited broadband must all be addressed. Ensuring digital access can help sustain a healthy agriculture sector that in turn can help rural areas stem themselves against depopulation, and help them retain young people. The concept sets out to create liveable villages, where people can and want to settle, because innovative, interconnected digital solutions improve their lifestyles. New business models, and platforms based on shared economy, currently concentrated in urban areas, are the best examples. However, this is just the beginning. The authors of this book described various technologies for completing this particular smart project, reflecting the many layers involved. We are hopeful that this approach helps you to grasp the complexity of this concept.

The smart villages concept has been proposed and successfully launched to a niche for the sake of rural areas, not only in the EU but also in the global arena as well. We had various discussions and consultations with different stakeholders including laymen and highly specialized academics and rural development practitioners. It has been pushed forward by our common efforts, by our persistent fight for the smart villages concept. We have been actively promoting this concept via a pilot project and preparatory actions since 2015. Indeed, we believe that smart villages offer the best way forward for a sustainable realization of the vision outlined in the Cork Declaration, while rural areas face a real and complex challenge. A challenge that needs to be tackled by a smart approach needs to pay exact attention to mitigate the digital divide between rural and urban areas and to develop the potential offered by connectivity and digitization of rural areas. Besides that, the need for integrated approaches and the complementarity and coherence interaction between different policy fields already emerged in this declaration. Throughout our common work, involving many publications, and the motivation of even more articles, we have always enjoyed inspiration from academics, from practitioners, and from a wide array of different stakeholders.

As generally perceived – and as we see it – this should be the role of Members of the European Parliament: representing the interest of the people, listening to their needs, translating, and further elaborating these needs with the help of the academics and practitioners. This assures that, at the end of the day, these new tools find their manifestation in (European) legislation. As politicians, we need to identify the needs that are important for future development. We must also acknowledge that many innovations here in Brussels are the results of input and feedback from our constituencies. Brussels has the potential to become a 'European Silicon Valley' not only at the legislative level. Read this book and you will become more familiar also with the process of idea-to-legislation practices.

Revitalizing rural communities and making them more attractive and sustainable is possible by using the full potential of information and communication technology. We believe that offering business opportunities, making rural areas more attractive for investors, and enabling farmers and other local actors to use their potential are the key to build successful rural communities. But the story cannot stop here; it is visible that more work needs to be done on this field and hopefully we will all have the opportunity to continue this useful and expected complex development approach.

We are personally satisfied, as over the past five years, our efforts as Members of the European Parliament were focused on building consensus around the necessity of rural areas and rural communities. We are grateful for the support we have received from several EU Commissioners in office, as well as from their Directorate-Generals, with our special thanks being extended to Mr Phil Hogan, Commissioner for Agriculture and Rural Development, and to the colleagues from DG AGRI.

We are equally grateful to all the outstanding authors of this book for their time and dedication, and especially to Dr Anna Visvizi, Dr Miltiadis D. Lytras and Mr György Mudri, the book Editors, for being the engines of this project!

Please read this excellent book, and join us in the effort to make rural areas truly successful again!

Tibor Szanyi and Franc Bogovic
Members of the European Parliament
Initiators of the Smart Villages' Projects in the European Parliament

Chapter 1

Smart Villages: Relevance, Approaches, Policymaking Implications

Anna Visvizi, Miltiadis D. Lytras and György Mudri

Introduction

Smart village may be a new, and for that matter a rather fancy, concept, yet the thrust of problems and challenges that it speaks to is by no means trivial or new. Therefore, the imperative inherent in the smart village concept and debate is to diagnose the status quo, propose viable ways of addressing problems and challenges, build consensus about the need to take action, and to actually follow the suit at micro-, mezzo-, and macro-levels. The truth is that villages — and so the romanticized picture of pastoral life and the beauty of nature, so close to childhood memories of many who will read the book — are under threat of disappearing. Several factors contribute to that. Perhaps the most important of them are progressing pace of urbanization coupled with the perceived opportunities that city life bears and negative demographic tendencies in rural areas. The problem is acute across the European Union (EU), but certainly, and regrettably so, it exists elsewhere too. As the process of depopulation of rural areas progresses, so is the heritage inherent in villages vanishing. At the same time risks, threats, and new challenges arise.

Rethinking the Rationale behind Smart Villages: Typology of Risks, Threats, and Challenges

As elaborated elsewhere (Visvizi & Lytras, 2018a), depopulating villages tend to be inhabited by the elderly, usually single, in need of medical care, help with cooking, and […] just company. These people tend to be deprived of the means for living and can hardly use electronic devices, should internet and electricity be available in their village. The nature of this challenge goes beyond the question of wellbeing and quality of life. It is in fact a question of life and death. Risks to their safety include the risk of not only burglary, assault, fire, and flooding but also malnourishment and sickness. Depopulating villages frequently lack basic

---

Smart Villages in the EU and Beyond, 1−12
© Anna Visvizi, Miltiadis D. Lytras and György Mudri, 2019
All rights of reproduction in any form reserved
doi:10.1108/978-1-78769-845-120191002

Table 1. Smart Villages: Typology of Challenges and the Corresponding Urgency of Action.

Temporal Dimension	Status of the Challenge	The Thrust of the Challenge	Prescribed Action	Type of Action
(1) Short-term	Emergency	Question of life and death, including safety and security	Action needed at this moment	What smart services, provided by whom, how, and at what cost, could be provided to ease the situation?
(2) Mid-term	Urgent	Question of wellbeing and quality of life	Planning and action needs to begin today – action needed today	
(3) Long-term	Very important	Question of cultural heritage, governability, and the cost of inaction	Planning needs to begin today – action needed in the near future	

Source: Visvizi and Lytras (2018a).

infrastructure, such as roads, reliable electricity grids, but also doctors, schools, and affordable groceries. This affects the wellbeing and quality of life in villages for its current inhabitants, both young and elderly. It also creates disincentives for possible newcomers and incentives for current inhabitants to leave. Depopulating villages frequently embody artifacts of inimitable cultural heritage, in terms of architecture, tradition (rites, habits), and oral history. Today's elderly are the guardians of traditions and heritage, but who will carry our shared tradition when its today's guardians are gone? There is also the big question of long-term implications of increasingly depopulated villages and rural areas, that is, the question of the state's administration ability to exert its control over those areas. These are real problems that require responses. Table 1 offers a snapshot view of these problems and challenges, while at the same time hinting to the urgency of action that needs to be taken to address them.

Smart Villages: The Concept and Its Relevance

The concept of smart village made its inroad into the policymaking and academic debates nearly simultaneously. On the policymaking front, the European Commission has been the champion of 'smart villages,' as reflected in Cork 2.0 Declaration of 2016 (European Union, 2016). The European Parliament upheld the momentum and so the Bled Declaration of 2018 adds further content to the idea and resultant policy strategies. Certainly, the concept and the idea have been borne out of several decades of debates and policymaking pertaining to mostly Common Agricultural Policy (CAP), Regional Policy, and then Cohesion Policy. In this context, the first Cork Declaration of 1996 (ECRD, 1996) should be pointed out. Today, the concept of smart village has been defined by the European Commission as follows:

> "Smart Villages are rural areas and communities which build on their existing strengths and assets as well as on developing new opportunities," where "traditional and new networks and services are enhanced by means of digital, telecommunication technologies, innovations and the better use of knowledge" (ENRD, 2018; European Commission, 2016)

With regard to the academic debate on smart villages, the rise of interest in the concept, its application, and implications can be related to the maturing of the debate on smart cities (Visvizi & Lytras, 2018a; Visvizi, Mazzucelli, & Lytras, 2017). It can also be related to the individual discovery process, that is, field research, personal experiences, and the resulting commitment to the cause of specific researchers. Research interest triggered by the sad examples of depopulating Greek villages in Peloponnese, especially in what used to be mythical Arcadia, or the equally tragically compelling example of Epirus, proves the point.

This volume brings these two rationales underpinning 'smart villages' together, thus making a case for a conceptually sound, academically disciplined, empirically driven discussion on smart villages and its relevance and application in the policymaking process. Indeed, the contributing authors represent both academia and the world of policymaking, thus confirming that dialog between these two is feasible and that synergies thus created can effectively feed the policymaking process.

The thrust of the concept of smart village that this volume advances is therefore delineated by three issues (see Figure 1 for details). First, the focus is the village seen as an ecosystem, a microcosm, a community, and people, rather than an aggregate, largely de-personalized construct such a 'rural area' or 'countryside.' Second, the value added by modern sophisticated information and communication technology (ICT) that can be applied in the context of a village is emphasized. Considering the two points aforementioned, by ontologically reifying the village, the focus of analysis in smart village research, as outlined in this volume, shifts to inhabitants of a given village, be it plural or individual (Visvizi & Lytras, 2018a). At the same time, a considerate and ethically and socially conscious use of ICT is advocated (Lytras & Visvizi, 2018; Visvizi et al., 2017). Notably, by focusing on specific needs and challenges individuals face, and by highlighting the value added by ICT, the so-defined debate on smart villages inadvertently engages itself with diagnosis of problems at hand and prescription of what needs to be done (Visvizi & Lytras, 2018b). This serves as a

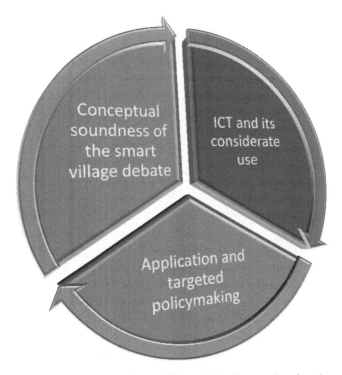

Figure 1. Smart Village: The Three Pillars of the Comprehensive Approach to Smart Village. *Source*: The authors.

bridge to the third segment of the approach that this book advances, that is, application and targeted, effective policymaking. Seen in this way, the approach to smart village that this volume promotes is comprehensive. That is, it brings together the specific to the academia conceptual zeal, highlights the value added by ICT, and engages in dialog with policymakers (Figure 1).

Smart Villages: The Imperative The Concept Entails

The imperative the concept and the debate on smart villages entail, to put it bluntly, is to save villages, their inhabitants, the heritage, the potential, as well as to preempt nascent risks. The challenge inherent in pursuit of this imperative is that the key stakeholders only rarely engage in a meaningful, constructive, and unbiased dialog. As Figure 2 outlines, comprehensive approach to smart villages, an approach that translates in targeted and effective policymaking, requires that all stakeholders are involved. This means that: village inhabitants are listened to and encouraged to execute their agency; the academia relates to the real problems and challenges and offers conceptually sound and methodologically disciplined frame in which these problems and challenges can be contextualized; that the think-tank sector recognizes the need to draw from

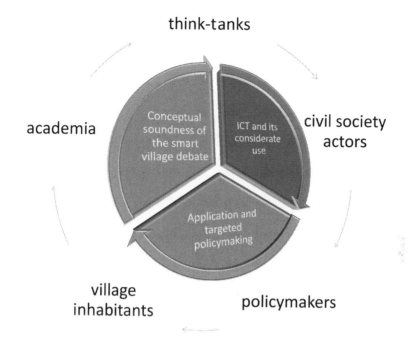

Figure 2. Smart Villages: From Needs to Targeted and Effective Policymaking. *Source*: The authors.

academic research and engages in dialog; the civil society actors are listened to carefully; and finally, a two-way bridge exists that links these actors with policymakers.

This volume proves that dialog engaging village inhabitants, the academia, the think-tank sector, civil society, and policymakers is possible. Representing the finest universities, think-tanks, advocacy groups, and the EU institutions, the authors contributing to this volume address diverse aspects of the debate on smart villages. As a result, from the policymaking point of view, this book offers not only a compelling insight into the key issues and challenges that the debate on smart villages revolves around but also a set of recommendations as to what can be done and what actions are already underway. From an academic point of view, this book defines the emerging field of research on smart villages. Departing from the research on smart cities, cognizant of the limitations and caveats specific to the latter (Visvizi & Lytras, 2018b), the comprehensive approach to the concept of smart village, that this volume advocates (Figure 1), invites inter- and multidisciplinary research focused on real needs and challenges and committed to devising ways of navigating them. ICT is seen as one of the means of so doing. At the heart of the smart villages' concept advocated here is the human being, inhabitant of the village, and his/her wellbeing. The chapters included in this volume skilfully draw from this conceptual and, indeed, normative framework, and convey hope that villages across the EU can be saved after all.

About This Volume and the Message It Conveys

This volume consists of 12 chapters addressing diverse issues related to the debate on smart villages. It offers a thorough insight into the concept of smart village, its application, and implications (Chapters 1, 2, and 3). The book is also filled with content drawing from developments in the field, including policymaking actions and strategies (Chapters 4, 5, and 6), issue and policy areas (Chapters 7, 8, and 9), and country case studies. Even if the focus of the volume is directed at the EU and the question of smart villages in the EU, a case is made that similar problems and challenges exist elsewhere. Chapters 10 and 11 elaborating on the cases of South Korea and the Gulf Cooperation Council (GCC) countries attest to that. Chapter 12 sheds light on the new research agenda and policymaking strategies that the smart village concept and debate necessitate.

In Chapter 2, titled 'Integrated Approach to Sustainable EU Smart Villages Policies,' Christiane Kirketerp de Viron and György Mudri elaborate on the origin of the smart village concept in the EU-level debate. As the authors explain, the concept of smart village emerged in EU-level policy debates on rural development in 2016, following the stakeholder-driven Cork 2.0 Declaration. It was developed through a pilot project initiative on 'Smart, Eco, Social Villages' and spelled out in the 'EU Action for Smart Villages' initiative. While the concept of smart villages remains unclear for many, substantial work has been carried out to develop the concept and to prepare the underlying supporting instruments at EU level over the last three years. The aim of this chapter is to give an overview of how the concept of smart villages has evolved at the EU level and to draw some recommendations for future policy work. The chapter reveals difficulties of the utilization efficiency of the EU funds in rural areas and shows a patched landscape of fragmented policy instruments. The key arguments are that while the mixture of these tools is important, the glue that binds them together is still missing and that the general utilization efficiency is not sufficient. The authors offer a set of five recommendations for the short to medium term, which is needed for the successful implementation of the smart approach: integration, simplification, communication, innovation, and 'rural proofing.'

In Chapter 3, titled 'Smart Villages Revisited: Conceptual Background and New Challenges at the Local Level,' Oskar Wolski and Marcin Wójcik argue that the smart villages' approach to rural development is promoted by the EU factors in the diversity of rural areas and the different nature of challenges faced by each area. The central role is assigned to local communities – formation of appropriate characteristics and attitudes that enable the creation of optimal conditions for development. This is also the result of the evolution of rural development policy, which is driven by the dynamics and direction of change of rural areas and changes in societal perception of change events in rural areas. The implementation of this development approach at the local level requires a transformation of the current school of thinking on development and the utilization of available resources. The key role in this process is played by local governments, which are part of the local community and also represent its interests.

The chapter combines theoretical and practical issues and represents a geographic perspective. Its first aim is to answer the question: How can local governments create the right conditions for smart development at the local level? The second aim is to discuss the smart village approach in the context of selected development concepts. This leads to a number of specific recommendations for policymakers. It also helps them to understand the approach, which is vital in the implementation of the aforesaid recommendations.

In Chapter 4, titled 'Toward a New Sustainable Development Model for Smart Villages,' Raquel Pérez-delHoyo and Higinio Mora explore the question of resilience in rural areas. As the authors explain, rural communities are increasingly open to a globalized world, and migration from rural areas to cities is becoming increasingly important. Many rural areas face depopulation, an aging population, and limited access to a range of services. To address this challenge, the ICT involved in the concept of smart villages have much to offer. In order to streamline the debate, this chapter proposes a methodology based on resilience. Resilience is defined as the ability of a habitat or system to recover to its initial state when the disturbance to which it has been subjected has ceased. In this regard, a retrospective of rural areas is proposed based on the experience of the garden city model, for which the advantages of rural areas were evident over those of urban areas. The objective is to reconsider the intrinsic qualities of rural areas in order to recover and enhance them with the added value of the EU Smart Villages approach. These facets will be the driving forces behind sustainable development. In conclusion, a number of recommendations are presented, including the development of a catalog, structured by regions and territories, of rural areas and their different potentials and opportunities, for the development of smart village projects.

In Chapter 5, titled 'The Role of LEADER in Smart Villages: An Opportunity to Reconnect with Rural Communities,' Enrique Nieto and Pedro Brosei focus on the program LEADER (Liaison Entre Actions de Développement de l'Économie Rurale", i.e., 'Links between the rural economy and development actions'), its evolution, and its value addition in the context of the debate on smart villages. The authors argue that over recent decades, rural areas have been facing significant challenges that exacerbate the existing discontent in their communities. These challenges are mostly reflected in depopulation trends, increased vulnerability to external shocks, and reduced quality of basic services. Local Action Groups (LAGs) all over Europe have been working on these challenges since the early 1990s. More recently, the smart villages concept is starting to generate enthusiasm among rural development stakeholders to try to revert these trends by supporting communities to move toward a more sustainable future while taking advantage of new, emerging opportunities. This chapter demonstrates that the LEADER approach and its principles are also part of the smart villages concept. However, practical differences between the two emerge as a result of limitations imposed by restrictive LEADER regulatory frameworks in many member states. Our main argument is that LEADER has what is needed to be the main tool for driving smart villages in Europe as long as there is a policy framework in place that enables LEADER to exploit its full potential. This conclusion is grounded on the analysis of the role that LEADER played in a number of smart villages initiatives across the EU.

In Chapter 6, Daniel Azevedo makes a powerful case for precision agriculture and its relevance for villages and their sustainability. As the author explains, stakeholders all over the EU, including academics, local and national authorities, business representatives, civil society, and the EU institutions are interested in creating a better life for citizens inhabiting rural areas. Building up on previous successful policies, including for instance the CAP and political actions, such as those outlined in the Cork 2.0 Declaration, these stakeholders assessed the current and future challenges of EU rural areas. The EU agri-food sector is not only the backbone of the EU rural areas but is also a driver of the EU economy, that is, delivering 44 million jobs in the EU and representing 3.7% of EU's gross domestic product (GDP). The EU is thus the world's number one exporter of agricultural and food products amounting to €138 billion in 2017. Today, the technological progress and the digital economy are transforming the EU economy in a way and speed never seen before; agriculture and food production are no exceptions. Obviously, the agriculture and the food production activities may have less obvious significance to the smart city but are key in the smart village concept. The ongoing technological transformation of agriculture will certainly enable and influence the design and implementation of any concept for smart villages. The question is therefore complex: How can agriculture and food production contribute and influence the smart villages concept and related policy strategies? This chapter dwells on this issue.

In Chapter 7, titled 'Energy Diversification and Self-sustainable Smart Villages,' James K. R. Watson explores the growth and opportunities of small-scale local power generation and the implications for internet access for rural communities. Solar power has grown exponentially in the last decade across the world and has provided opportunities for the development of local energy communities and on microgrids across the world and in Europe. The huge cost reductions experienced in generating solar power and its relative mobile and flexible nature have made it a technology perfect for rural areas to develop their own sustainable source of electricity supply. The increasing rise of digital tools has coupled nicely with the advent of the mass use of solar power in rural areas and thus the connection between smart solar and smart villages has become increasingly a norm. Rural communities in Europe have embraced solar technology, with many farmers using solar power as a means to reduce their electricity costs and also generate new streams of income to improve their overall livelihoods. Some case studies from India, Germany, and Africa will be examined. Other experiences will also be considered, especially where double land use between solar power and livestock has empowered rural communities. Outside of Europe, Africa and Asia have also seen solar power as a means to electrify remote rural villages. This has led to the development of microgrids and new technologies that are less deployed in Europe, which are being rolled out for rural communities in the rest of the world. This has been particularly successful in creating smart rural communities, as often digital communications have already reached these communities and thus power and telecoms are combining to provide clean and controlled power for millions in Africa. This chapter will also assess the growth of smart energy communities in non-traditional energy markets and determine what lessons we can learn from their experiences. This chapter examines other sources of renewable energy and the role that biogas, biomass, and others are playing in the creation of smart villages in Europe and beyond. Biomass has been the traditional tool for many rural communities to generate power and heat, and thus an examination of how it now plays a role in smart villages is vital to understanding the energy transition we are experiencing in rural communities.

In Chapter 8, Xenia Szanyi-Gyenes explores the role of smart and medium-sized enterprises in the smart villages concept. As the author explains, to create an operative smart system of smart villages, it needs the participation of small- and medium-sized enterprises (SME). In the smart villages concept, local needs require real local solutions; a kind of 'I can do this for you' philosophy. SMEs, especially, micro-enterprises, or even, self-employed individuals, have the potential and the capacity to develop local solutions to local problems and have the flexibility to think on a very micro-level. New ideas are needed for smart villages, including new solutions and new perspectives. The potential of success is in the SMEs, indeed. Because it is not enough to create a system, it must be operated too. Small businesses can ensure the effective functioning of smart villages. The idea of smart villages is about people. It is intended that the rural population be able to use all modern technological tools and get closer to the services common in an urban environment. The question is how to make rural life attractive, especially for the young generations. To this end, we need smooth connections by broadband internet and enhanced potential for mobility. It is also a social and ecological project that is driven by public efforts assisted by larger budgetary means or, in the case of the EU, by a good coordination of the various development funds with broader rural development goals. However, we should not believe that SMEs operating in small settlements are to become more competitive than those in big business hubs. Its needed to acknowledge that matching urban/rural balances is a matter of financial solidarity, and thus we can keep our landscapes soundly populated and protected.

In Chapter 9, titled 'Smart Villages in Slovenia: Examples of Good Pilot Practices,' Veronika Zavratnik, Andrej Kos, and Emilija Stojmenova Duh contextualize smart rural development in Slovenia. To do so, the authors present the current state of the art, elaborate on new interdisciplinary approaches in understanding the problems of rural areas, and apply new approaches to smart rural development. The chapter describes the state of the art in Slovenia in connection with smart and sustainable rural development and examines the results. It looks at the solutions and practices that local communities found to answer the existing challenges. Three promising examples are presented, related to different sectors: tourism, mobility, and innovation. The last part of the chapter contextualizes our findings and further explains our approach to smart rural development. Finally, the chapter introduces a new concept – that is, Smart Fab Village. The concept of Fab Village is built on the concept designed for urban areas and then carefully adapted to the needs and specific requirements of rural areas.

In Chapter 10, titled 'Smart Village Projects in Korea: Rural Tourism, 6th Industrialization, and Smart Farming,' Jonghoon Park and Seongwoo Lee investigate diverse policy experiences of smart village strategy in Korea. The Korean approach has been highly influenced by the EU experience emphasizing the importance of a bottom-up territorial development. The Korean government acknowledges agriculture is not the only driver of rural jobs and wealth creation. Rather it understands diversified non-farm activities in rural areas are essential to revitalize the rural economy. The major policies relevant to the development of rural smart village are, first, establishing a regional innovation system fitted for depressed regions, second, inducing agriculture to become value-added industries, third, diversifying rural economic activities and integrating industrial support, fourth, improving the welfare of rural residents by improving settlement conditions, and finally, encouraging rural–urban interaction. Since the campaign of smart rural village as a rural development strategy is closely related with the discussion of rural tourism in Korea, this study investigates past and recent streams of rural-tourism strategies pursued by the central government in Korea. Along with introducing the historical development strategy in Korea, this study presents the current and possible future characteristics of rural development strategies in Korea. This study investigates the perceived role of tourism as well as recent streams of rural development policies such as 6th industrialization and smart farming in the rural development strategies. Presenting success and failure stories, this study also considers why development of rural tourism has been slow in rural areas in Korea, reviewing restraints, reservations, and problems identified during the last decades in Korea.

In Chapter 11, titled 'Smart Villages and the GCC Countries: Policies, Strategies, and Implications,' Tayeb Brahimi and Benaouda Bensaid explore the notion of depopulation of rural areas and ways of addressing the resulting challenges in the GCC countries. As the authors highlight, the majority of the population in the GCC lives in urban areas. By the year 2030, the percentage of the rural population is expected to be 14% in the Kingdom of Saudi Arabia, 11% in the United Arab Emirates, 10% in Oman, 9% in Bahrain, 1% in Qatar, and will remain 0% in Kuwait. Like many other countries, however, GCC countries continue to invest efforts and resources toward their national agendas aimed at sustainable development. In this context, smart villages are of special interest as, increasingly, they serve as the crossroads between urban living and rural life embracing history, culture, tradition, spiritual values, and human capital. The objective of this chapter is to explore actions taken toward the development of smart villages in the GCC countries, with a comparative overview on pertaining approaches and strategies; challenges related to the implementation of these actions are identified. It is demonstrated that despite GCC's tremendous efforts toward developing infrastructure in urban centers, more infrastructure investment is needed with regard to key issues related to developing remote areas – including their smart networks, digital facilities, and e-governance. It is also highlighted that more research is needed, especially on issues related to the transformation of villages into smart villages, including the need for holistic approaches, policies, and strategies toward smart villages. In Chapter 12, Miltadis D. Lytras, Anna Visvizi, and György Mudri bring the key points and observations made in this volume together to highlight the future research and policymaking avenues in the field.

Final Remarks

Overall, the chapters included in this volume offer a comprehensive take on the debate on smart villages as it unfolds in the EU and elsewhere. By examining the concept of smart village from a broad inter- and multidisciplinary perspective, the chapters included in the volume populate a broad field defined by not only politics, political science, and economics, but also sociology, geography, regional studies, and entrepreneurship. This volume is the first ever publication that addresses explicitly the concept of smart village, including the conceptual framework upon which it is built, current developments, and policymaking considerations. The value-added of this volume can be derived from the following observations. The book offers an overview of the academic debate and policy strategies associated with the concept and debate on smart villages in the EU and in other parts of the world. The authors contributing to this volume, including academics and practitioners, provide detailed accounts of diverse issues pertinent to the smart villages debate. The case studies and real-life examples that have been employed in the discussion mirror the developments shaping the debate on smart villages and resultant policymaking in the EU and international contexts. As a result, the volume entails an integrated discussion on diverse interconnected aspects of the smart village debate. Written in an approachable way, this volume will be of interest to academics, researchers, and practitioners.

This volume brings together and discusses critically the well-established, emerging, and nascent concerns and questions related to the challenge of depopulation of rural areas and the implications of this process. The promise and value-added of ICT to address the problem at hand is queried. By embedding the discussion in a broad well-founded conceptual framework and reaching out to case studies, this volume offers a fascinating journey across issues and areas pertaining to the concept and, indeed, policy field of smart village. By integrating views and insights from leading experts, academics, and practitioners in the field, as well as by focusing on several issue areas, this edited volume will serve as a primer in the discussion on smart villages and ways of navigating the plethora of challenges related to it.

References

ECRD. (1996). Cork Declaration – "A living countryside," European Conference on Rural Development (ECRD), Cork, Ireland.

ENRD. (2018). Smart villages portal. European Network for Rural Development (ENRD). Retrieved from https://enrd.ec.europa.eu/smart-and-competitive-rural-areas/smart-villages/smart-villages-portal_en

European Commission. (2016). EU action plan for smart villages. Brussels: European Commission. Retrieved from https://ec.europa.eu/agriculture/sites/agriculture/files/rural-development-2014-2020/looking-ahead/rur-dev-small-villages_en.pdf

European Union. (2016). Cork 2.0 Declaration – "A better life in rural areas," Luxembourg: European Union. ISBN 978-92-79-63528-1. Retrieved from https://doi.org/10.2762/370418KF-01-16-997-EN-N

Lytras, M. D., & Visvizi, A. (2018). Who uses smart city services and what to make of it: Toward interdisciplinary smart cities research. *Sustainability*, *10*(6), 1998. doi:10.3390/su10061998

Visvizi, A., & Lytras, M. D. (2018a). It's not a fad: Smart cities and smart villages research in European and global contexts. *Sustainability*, *10*(8), 2727. doi:10.3390/su10082727

Visvizi, A., & Lytras, M. D. (2018b). Rescaling and refocusing smart cities research: From mega cities to smart villages. *Journal of Science and Technology Policy Management*, *9*(2), 134–145. doi:10.1108/JSTPM-02-2018-0020

Visvizi, A., Mazzucelli, C. G., & Lytras, M. (2017). Irregular migratory flows: Towards an ICTs' enabled integrated framework for resilient urban systems. *Journal of Science and Technology Policy Management*, *8*(2), 227–242. doi:10.1108/JSTPM-05-2017-0020

Chapter 2

Integrated Approach to Sustainable EU Smart Villages Policies

Christiane Kirketerp de Viron and György Mudri

Introduction

Rural and intermediate areas account for 91% of the territory of the European Union (EU). It provides 43% of the EU's gross value added, and are home to 60% of its population (Committee of Regions, 2017a). In spite this, rural settlements are lagging behind compared to cities, as shown by a number of economic and socioeconomic indicators including broadband coverage where 80% of urban population has access to next-generation networks against only 47% of rural households (European Commission, 2017a).

The Committee of Regions (CoR) also showed that despite this disparity, EU funding for non-agricultural activities in rural areas decreased in the EU financial period 2014–2020 compared to previous periods. This is also reflected in the current debate around rural development which reveals concerns that underinvestment in the rural economies and in rural services will contribute to widening the gap. The speed at which the economy and society is embracing digitalization can only acerbate the situation in so far that rural areas are not partaking in the digital economy. Increasingly, this disconnect between rural and urban, center and periphery are considered to spill over into voter behavior, with rural areas being at higher risk of populism and more prone to euroskepticism (ECFR, 2016).

The concept of smart villages is conceived also as a policy objective, that is, a way of connecting rural communities to the digital economy and help rural communities overcome issues related to low service provision, lack of high-value jobs, difficult logistics for industry, and a number of other issues caused by the remoteness. Preliminary fact finding work carried out by the European Network for Rural Development (ENRD), which shows that smart villages can emerge as the fruit of cooperation and association between villages, as well as between villages and nearby urban areas. A number of cases have also shown that it is possible to develop into a smart village as a single entity without cooperating. A key ingredient across the board, however, is the involvement of the local community as well as the use of digital tools.

The intricacies and specificity of small rural settlements in the EU require a different policy analysis. The objective of this chapter is to map the relevant EU policies and discuss pathways for improving the effectiveness and efficiency of these policies toward smart villages. The size of the available funds is not always the main question for smart rural, regional, or village development, but the actual approaches of the different funds. The approach can make advantages also from the already operational tools, such as the "smart cities" (Sedov, Chelyshkov, Rujitskaya, & Solntseva, 2016).

The aim of this chapter is to examine the possible integrated approach of the different funds affecting the rural areas and to different pathways to develop a more integrated approach for smart villages at the EU level. The focus is not only on the financial funds but also on the underlying philosophies of the policies. Smartness of a settlement is composed of several dimensions; the aim of this work is to identify the main measurable elements from the policymaking side.

There are possible ways to integrate the rural development related policies by analyzing the most important governance and policy aspects of the existing policy tools. This exercise goes further than the reflection in the Cork 2.0 Declaration, which is clearly calling for a common strategic and programming framework for the Common Agricultural Policy (CAP) as a whole. For smart villages to be successful, a common framework for rural areas may be needed.

To this end, the argument in this chapter is structured as follows: (1) description of the policy discussions relating to smart villages at the EU level, (2) overview of EU policies active in the field of smart villages, (3) institutional and stakeholder-driven recommendations, declarations, and opinions in line with the integrated policies, and (4) recommendations for short- and medium-term actions to policymakers and academia to further promote and develop smart villages through EU policies.

EU Policies for Smart Villages

What Are 'Smart Villages' in the Present Policy Discussion?

A novelty in the European policy discussions, the concept of EU smart villages was developed as a cooperation between several Commissioners and their Directorate-Generals as well as the European Parliament (EP), starting in 2015. Although the European Commission remains the primus motor, the concept has been enhanced and enriched through the work of other EU institutions such as EP, CoR, and the European Economic and Social Committee (EESC) through a number of initiatives, opinions, events, pilot project, preparatory actions, and working groups.

As in the case of smart cities, digitalization remains a focal point of the concept of smart villages, but it is not an end in itself. Smart villages focus more widely on economic sectors, social aspects, innovation, and other policy fields (Visvizi & Lytras, 2018a). By addressing issues holistically, the aim is to make rural areas attractive for people to live and work and allow innovative and digital solutions to improve their life quality. This comprehensive approach to the variety of issues and problems at hand has been elaborated in detail in the most recent literature on the subject (Visvizi & Lytras, 2018a, 2018b; Visvizi, Lytras, Damiani, & Mathkour, 2018).

The first document published by the Commission to this end, the 2017 'EU Action Plan for Smart Villages,' developed a definition of smart villages which to a certain extent reflect the concept of smart cities. Accordingly:

> the concept of smart villages refers to rural areas and communities, which build on their existing strengths and assets as well as on developing new opportunities. In smart villages traditional and new networks and services are enhanced by means of digital, telecommunication technologies, innovations and the better use of knowledge, for the benefit of inhabitants and businesses. (European Commission, 2017b)

The emergence of the notion of smart villages is also closely associated with the 'CORK 2.0 Declaration for a Better Life in Rural Areas' (European Commission, 2016) which was issued by rural stakeholders in 2016.

Since the publication of the 'Action document,' a specific Thematic Working Group and a Smart Villages Portal has been set up under the ENRD (European Network for Rural Development, 2017a), and the concept of smart village paved its way in academic research. The Bled Declaration on smart villages (Bled Declaration, 2018) was adopted in April 2018 with the contribution of the EP and four Directorate-General (DGs) of the European Commission: the Agriculture and Rural Development (AGRI), the Regional and Urban Policy (REGIO), the Mobility and Transport (MOVE), and the Communication Networks, Content and Technology (CNCT), while a high-level event in Gödöllő, Hungary in October 2018 added the education, skills, and the energy aspects to the smart villages concept through the participation of the Directorate-General for Education, Youth Sport and Culture (EAC) and the Directorate-General for Energy (ENER).

An EP-sponsored pilot project on "Smart Eco Social Villages" (European Commission, 2017c) contributes to developing the knowledge base for smart villages. This will be followed up by two preparatory actions which are currently in different stages of preparation. Within the EP, the support for smart villages has gradually grown across parties and member states and now has the support from Hungarian, Slovenian, French, German, Irish, Italian and Slovakian Members of the European Parliament (MEPs), representing the Progressive Alliance of Socialists and Democrats (S&D), the European People's Party (EPP), and the Greens/European Free Alliance (Greens/EFA) groups.

It should, however, be noted that the document 'EU Action for Smart Villages' was not a formalized publication of the European Commission. That is, it was not subject to a decision of the College of Commissioners. It therefore has no legally binding value. Rather, it should be seen as a communication product, or indeed a joint statement of intent by three Commissioners.

While a number of Directorate-Generals from the European Commission are involved in the inter-service discussions on EU smart villages, the concept only appears explicitly in one sector proposal related to the new Multifinancial Framework – that is, the "Proposal for Regulation on CAP Strategic Plans, COM(2018)392" (European Commission, 2018a). Here a specific result indicator has been developed for smart villages. The concept also gets a mention in two recitals – one which refers to the aims and objectives of the CAP and one which establishes the scope and potential for types of support under the cooperation type of intervention. From this, one can derive that it is the understanding of the Commission that the smart villages concept should be rolled through models of cooperation rather than as a top-down approach and that member states should consider the potential for developing smart villages when carrying out the strength, weaknesses, opportunities and threats (SWOT) analysis and developing their CAP strategic plans. Apart from that, smart villages do not appear explicitly in other future policy plans within the EU.

Evidently, 'smartness' as such is not new in the European context. It appears in many other acts and plans, such as the Europe2020, the concept of Smart Specialisation under Cohesion policy (European Commission, 2011), or the Digital Single Market (European Commission, 2015a). The concept of smart cities was first coined systematically in the EU in 2007 and has been integrated into a number of EU policies such as Regional Development and Research and Innovation. There has, however, been few linkages established at present between the concept of smart cities and that of smart villages despite the fact that a lot of work is being done on rural–urban linkages.

It is clear from the aforementioned that the interest in developing smart villages is showing an upward curve in EU policymaking. Yet the concept remains diffuse, hard to capture, and difficult to nail down in a one-size-fits-all definition, even within a given geographical context. This can hardly be surprising given the experience with defining smart cities, which also proves cumbersome to pinpoint and which in the case of the United Kingdom has been proposed as a process rather than a product (DBIS, 2013).

What can be concluded about the EU approach to smart villages is that it remains open-ended to allow for different aspirations and avoid limiting opportunities and innovation capacity. How the wider EU policy landscape can cater for smart villages will be the role of the introductory steps and the analysis of the possible deeper integration of the concept in all European Structural and Investment Funds (ESIF), such as the European Agricultural Fund for Rural Development (EAFRD), the European Regional Development Fund (ERDF), the European Social Fund (ESF), the Cohesion Fund (CF) and the European Maritime and Fisheries Fund (EMFF). The links to other instruments, such as the Research and Innovation Framework, will be dealt with in the next subchapters.

EU Policy Framework for Smart Villages

The Common Agricultural Policy

The CAP remains the largest single policy of the EU in budgetary terms. It currently consists of two pillars. The first pillar with direct support for farmers (income support) and market measures account for 75.6% of the 408.313 billion which is allocated to the CAP for the 2014–2020 period. The second pillar, the Rural Development Policy, accounts for 24.4% of the CAP funding. In contrast to the first pillar, rural development is co-funded by member states. The original EU allocation of approximately 100 billion euro to the rural development therefore generates a total budget for rural development of 151.8 billion euros, where member states' contributions are taken into account.

Despite its name, it should, however, be noted that the Rural Development Policy for the largest part is focused on agriculture – a notion which is perhaps better represented in the fact that the underlying fund is called European Agricultural Fund for Rural Development. Approximately 85% of the rural development funds are used for various investments, knowledge and environment-climate schemes for the farming sector, with a small share of this going to develop the food and forestry sectors. In reality, only 23.3 billion euros of the total public expenditure is planned for developing other aspects of the rural economy as well as improving the provision of services in rural areas (priority 6 of the Rural Development Policy).

There are historical explanations for this. Rural development was not on the drawing board at the time of the establishment of the CAP. In the aftermath of the Second World War, the main focus was on improving the agricultural situation in the member states; food security was a key concern and the rural economy was to a large degree still agricultural.

This is also reflected in the treaty articles that concern the CAP. Article 39 of the Treaty on the Functioning of the EU (TFEU) sets out the specific objectives of the CAP: increase agricultural productivity by promoting technical progress and ensuring the optimum use of the factors of production, in particular labor; ensure a fair standard of living for farmers; stabilize markets; ensure the availability of supplies; and ensure reasonable prices for consumers (TFEU, 2007).

Since the first steps in 1988 toward establishing a rural policy, the EU approach to rural development has been quite consistent (Matthews, 2007). The three overall objectives – structural adjustment of farming, environmental protection, and social cohesion – have only been changed by name and not nature throughout the years.

Nevertheless, the first Cork Conference in 1996 marked the beginning of the new approach to the rural aspects of Rural Development in the EU. The Cork Declaration, which fed into the preparatory work for Agenda 2000, called for an integrated, territorial approach led by bottom-up initiatives best exemplified by the LEADER approach. With Agenda 2000, national and regional rural development programmes became the *ordre du jour*, bringing together all non-market structural measures from the CAP in a programmed framework.

Broadband was added to the priority area for action in the 2008 Health Check and the 2013 CAP reform introduced a stronger notion of cooperation and coherence between EU funds in rural areas. Rural Development Policy became subject to a double set of rules, the Common Provision regulation governing all ESIF and the rural development regulation, setting out the fund-specific rules. This duality of rules is considered to have caused issues with administrative burden. This was also one of the conclusions in the High Level Group on Simplification for ESIF.

Throughout the years, there has been a recurrent tension between the agricultural aspects and the territorial approach to rural development. This has also spilled over into the policy debate with a recurrent question of where rural aspects are best managed — as part of the CAP or as part of Regional Development Policy.

Within the current 2014–2020 Rural Development Framework, the range of measures offers a broad toolbox for smart villages. This includes measures on the knowledge transfer and information actions, investment in physical assets, farm and business development, basic services and village renewal, cooperation, and the LEADER/community-led Local development (CLLD) measure.

These measures on the list can be considered as opportunities for smart villages; however, in an alone-standing capacity it is hard to see how a single measure could cover all aspects needed to develop a smart village. The two most apt measures would be the LEADER/CLLD with its bottom-up local development strategy under which smart villages could be an objective, and the cooperation measure which in fact is very open and has the capacity to fund most types of cooperation that contribute to the objectives of the Rural Development Policy. This measure is also used for finding the European Innovation Partnership for agricultural productivity and sustainability (EIP-AGRI), which supports innovative projects related to farming and forestry and which functions as a bridge, plugging knowledge from the Research and Innovation Framework into initiatives promoted or supported under the CAP (European Commission, 2012).

For both LEADER/CLLD and the EIP-AGRI, the driving principle is to trigger innovation and development through cocreation and involvement of end users. However, whereas LEADER is local, territorial, and community-driven, the EIP-AGRI is sectorial, works across borders, and has a strong capacity to forge links to research and technological developments. There is also a stronger degree of top-down agenda setting in the EIP-AGRI. Nevertheless, both methods bring something to the table in terms of developing smart villages. It highlights the need for flexibility and subsidiarity, the need to respect territorial sensitivities, the implementing mixture of top-down and bottom-up tools, the need for strong local anchoring, and the need for central top-down measures to promote digital and technological rollout.

The authors observe that this notion of flexibility, adaptability, and versatility is also reflected in the new CAP philosophy. This is particularly the case for the typical rural development support where the Commission proposal moves away from predefined measures with established eligibility criteria in order to allow for member states to have more room for maneuvre in designing CAP interventions.

This approach seems to be welcomed by most member states; however, experience has also shown that coordination, facilitation, and central 'knowledge hub' support from the Commission improves the development of support schemes. Therefore, this type of support should be offered at the EU level in order to ensure that member states develop the best possible schemes to promote the development of smart villages.

Cohesion Policy (CF, ERDF, ESF)

Cohesion policy is the other big spender of the EU budget. Underpinned by three funds, it aims to promote economic, social, and territorial cohesion by reducing disparities between the various regions and the backwardness of the least-favoured regions Among the regions concerned, particular attention shall be paid to rural areas, areas affected by industrial transition, and regions which suffer from severe and permanent natural or demographic handicaps such as the northernmost regions with very low population density and island, cross-border, and mountain regions (TFEU Article 174, 2008).

Despite this treaty reference to rural areas, a CoR study has found that the ERDF is increasingly financing urban areas. Only 25.8% of the resources allocated under the ERDF have been granted to rural areas in the previous programming period. This should be seen against a backdrop of a deepening of the gap between rural and urban in terms of infrastructure or the digital services and skills (Committee of Regions, 2017a).

This trend is supported by the 'smart cities' which is operating under the ERDF. Nevertheless, there are lessons to be learned for smart villages from this initiative – notably the lack of one-stop shops for (funding) services, single access points to network services, and Big Data analytics for advances decision-making – but definitely this aspect can be further improved in rural areas (Lytras & Visvizi, 2018).

One of the areas where the structural funds are particularly apt is broadband. However, uptake has been modest and a lack of capacity in national/regional administrations has been identified as a factor preventing faster rollout. In 2018, the number of projects selected corresponded to only 1/5 of the planned allocations. A specific project under interregional cooperation (INTERREG) has designed methodology which help communities develop new ideas for rural services and businesses linked to the digitalisation.

As for the EU skills agenda, according to the CoR, the European Social Fund is also struggling to deploy its vocational training resources in rural areas: only 7% of the ESF is dedicated to rural areas in the current programming period (Committee of Regions, 2017a). Yet there is a direct correlation between the youth employment rate and the percentage of young people in vocational training.

The authors believe that this share does not reflect the territorial and population weight of the rural areas. They also believe that besides the amount, the synergies also play a major role in the efficiency of these funds.

There are grounds for optimism. The Commission proposal for the future regional funds includes a 'priority objective' on mobility and ICT connectivity, with a 'specific objective' on enhancing digital connectivity. Corresponding to this specific objective, there is an 'enabling condition,' which requests having a regional or national broadband plan, assessing investment gaps, presenting a planned intervention that enhances access to affordable future-proof infrastructure, etc.

The proposal includes another priority objective, "sustainable development of urban, rural and coastal areas and local initiatives." The combination of rural–urban with a strong local anchoring can create a favorable condition for supporting smart villages, notably by streamlining and simplifying territorial tools for rural communities.

Research and Innovation

Under the current Research and Innovation Framework, Horizon 2020, rural research is often associated with the "Societal Challenge 2: Food Security, Sustainable Agriculture and Forestry, Marine Maritime and Inland Water Research and the Bioeconomy."

As the title reveals, within this societal challenge the main focus is on the four Fs: food, fish, farms, and forestry. Nevertheless, there is also a number of projects relevant for the wider development of rural economies and services. These projects focus, for example, on social innovation (SIMRA), on better rural innovation through networks (LIAISON), on the development of living labs in rural areas in order to diversify the rural economy (LIVREUR), on new rural business models (RUBIZMO) and on identifying and promoting policies, governance models, and practices that foster mutually beneficial relations between rural and urban areas (ROBUST). Six projects funded under Marie-Skłodowska Curie Actions worth almost 5 million euro of additional EU contribution are also relevant, including for example research and innovation staff exchange on social innovation in rural areas and on social entrepreneurship in structurally weak regions (RURACTION). Digital Innovation Hubs have also appeared on the horizon in recent years

An often-overlooked aspect is the fact that there is a wealth of excellent research being carried out, which may not be labeled 'rural' as such but which has significant potential to fuel innovative solutions for rural areas. This includes the work carried out related to smart cities, projects related to the development of renewable energy, carbon neutral solutions, e-health, and automated vehicles. All of this research is relevant for rural development as it all brings solutions for typical rural issues – for instance automated on-demand vehicles may provide the solution to the long-standing issue of rural mobility and lack of public transport.

Another relevant area under the strong development is the bioeconomy. Historically, the bioeconomy has been perceived as a research topic; however, recent developments have made clear that the bioeconomy is ready to be deployed in the European economy. Experience from big demonstration plants such as AgriChemWay in Ireland and Novamonte in Sardinia funded under the current research and innovation program shows that a sustainable and inclusive bioeconomy has the potential to be a real game-changer for rural economies, creating around a million new jobs in rural and coastal areas, many of them with high value (European Commission, 2018b). Investments and innovative business models are needed, as well as a mix of local anchoring and top-down support. Already today there are around 3,000 renewable energy cooperatives in the EU, with most of them concentrating in the northern member states. There are grounds to believe that this trend can easily be mainstreamed in the future.

There is no shortage of research and knowledge being developed; however, finding ways of better tapping into this knowledge, of channeling ideas out of the labs and onto the ground, remain a key challenge. Whereas the EIP-AGRI was conceived to provide this kind of bridging function for agri-food and forestry research, no such tool exists for rural development when it comes to mobility, health, education, and the jobs of tomorrow. A new role for the successor of the ENRD could be a potential solution but it would require a stronger interaction with the world of research, entrepreneurship, and financing.

Stakeholders' Positions to Improve Coherence of EU Policies for Rural Areas

A number of associations and organizations have come forward with ideas for how to improve the coherence of the EU policies for rural areas. The following section presents some of these ideas, which may also be relevant for smart villages, and considers whether these have been taken up in the proposals for the future.

The Cork 2.0 declaration "A Better Life in Rural Areas" was the product of a conference in which more than 350 rural representatives, stakeholders, and academics discussed the aspirations of rural communities and the policies needed to underpin such ambitions. The declaration sets out 10 policy orientations, of which three refer to how to improve the efficiency and effectiveness of policies active in rural areas. Many of these recommendations, such as the call for the CAP to be based on a single strategic and programming framework that provides for targeting all interventions to well-defined economic, social, and environmental objectives or the call for subsidiarity and strengthening of locally led initiatives, are to various degrees reflected in the Commission proposals for a CAP strategic plan.

The declaration also calls for the development of a rural proofing mechanism, which would systematically review other macro and sectoral policies through a rural lens. The notion of rural proofing builds on the experience of rural proofing in Canada as well as the United Kingdom. The OECD has also carried out important work in this field and it has been subject to a great deal of attention from the side of academia (Sherry, Erin & Shortall, Sally, 2018; Shortall, Sally & Alston, Margaret, 2016). Rural proofing asks of policymakers to not only consider the impact of a given policy initiative on rural areas but also to consider specific needs of rural areas when preparing proposals. Within the European Commission's Better Regulation toolbox, there is still no 'rural proofing' tool although there is a 'Territorial Tool.' The latter is however not used systematically. (European Commission, 2015b)

The interest organization Ruralité, Environnement, Développement and the European Countryside Movement have also developed policy ideas. Backed by the CoR, the rural movement is calling for the development of an EU Rural Agenda which would integrate elements from several policies and create a European Multifund for rural territories. This could be seen as a pendant to the EU Urban Agenda (R.E.D., 2017).

This request was also reiterated in the 2017 Venhorst Declaration, which was penned by the European Rural Parliament, an umbrella organization of LAGs and other local stakeholders. The blueprint went a step further asked for a minimum contribution of 10% from each European Structural and Investment Fund, to be allocated to CLLD/LEADER.

Within the Commission there seemed little appetite for the establishment of a new fund and the request was not picked up when the Commission developed its proposals for the next Multifinancial Framework, nor in the sectorial policies (European Commission, 2018c). This could be explained by the strong presence of incrementalism in the EU budgetary process, where each DG operates like Lindblom's original governmental agency watching over specific sectoral interest and associated budgets (Lindblom, 1959). In such a scenario, the rural interest, being territorial, has no such governmental agency to represent it and will lose out.

Meanwhile, the opinion of the EESC (European Economic and Social Committee, 2017) and the opinion of the CoR further defines the smart villages approach. The chapter emphasizes that for the sake of a sustainable and effective program for smart villages, there is a need for a better coordination and synergies between the relevant EU policies and funding streams. Also, the development of an integrated policy and support instrument for smart villages is needed. The EESC also proposed to introduce innovation brokers in rural areas – people who are changed with supporting, developing, and finding appropriate finance for innovative projects.

It could be argued that the fact that the EAFRD is no longer covered by the Common Provisions Regulation is a signal that rural development is taking the more sectorial road rather than territorial. Whereas this increases consistency and coherence with agricultural aspects of the CAP, there are concerns that it may have a negative effect on the general consistency of support for the wider rural economy. Nevertheless, the overall future orientation of the CAP seems to suggest a closer connection to the territory in which it is being implemented, with far more flexibility and subsidiarity for member states (MS) to design rules that fit with other EU and national funds.

Five Ways Forward for Smart Villages

This chapter paints a picture of a European policy landscape with many policies and instruments that are relevant for smart villages, however fragmented and underutilized. So far, the concept of smart villages has only been firmly anchored into the proposal for the future framework of the CAP, illustrated by the presence of a smart villages result indicator as well as a mention in two recitals. With the current Commission entering the last phase of its mandate, new initiatives will not be forthcoming. It is therefore important to seek to new develop the smart villages concept within the current structures. The authors propose five avenues to take.

Integration

It is clear that if the concept of smart village is to gain any traction and have a large-scale impact, it cannot remain confined to the CAP. However well-healed, this policy alone does not have the scope, scale, and means needed to deliver on smart villages throughout the EU.

Connectivity remains the Achilles heel of any smart village strategy. The EU is providing a plethora of different funding opportunities with specific national and EU level broadband offices to deliver. In 2018, the Commission published a five-legged action plan for rural broadband but there is a need to streamline, strengthen, and develop initiatives related to the other two pillars of digital transformation: digital skills and services. In this context, it is also worth further exploring the better use of financial instruments as it is done in relation to the implementation of Urban Agenda.

At the local level, success for smart villages starts with having a plan. A pertinent question to ask is whether the same would not be applicable at the European level. Various pilot projects and preparatory actions can give insights into how a smart village can develop, with recipes for how to mix and match EU funding and national instruments. The authors are convinced that without a coherent EU level strategy and the further development of knowledge hub capacities, the system will neither be effective nor efficient. More work is needed to ensure that this happens seamlessly.

Simplification

Bureaucracy is still the number one enemy of innovation, but in the era of e-governance and burgeoning artificial intelligence, it should not be considered farfetched to say that better coordination and simpler implementation starts already at the European level. One cannot expect project managers to successfully navigate four or five different rule sets, each with different application forms and procedures. The need for one-stop shops for funding remain as relevant as ever.

The analysis reveals lack of simplification and systematic difficulties of the utilization efficiency of the EU funds in rural areas, even if some level of integration is given, such as in the case of the CLLD. While the mixture of these tools is important, the utilization efficiency is not sufficient regardless of the top-down or bottom-up implementation. Centralized support for the development of simple schemes and measures is crucial to improve the mutual learning and improve the administrative capacity of and in member states.

Communication

Generally speaking, communication remains a soft spot in the European project, with several news commentators linking euro-skepticism and populism in rural areas as a direct consequence of it. Understandably, EU citizens tend to be unaware of the complexity and interactions of the EU instruments, and rightly so. Only a thin layer of the society is in direct contact with EU funding and even fewer with rural development funding. In the current period, a total of around 2,600 LEADER Local Action Groups may cover 54% of the rural population but that does not necessarily mean that citizens are aware or even affected.

Against this backdrop, smart villages have the potential to become lighthouses – real-life examples of how the EU supports the positive development of rural communities. One way forward would be to have this type of demonstrators – living labs – in place throughout the Union. This would serve the double purpose of not only communicating with citizens but also helping local communities and authorities see the potential in becoming a smart village. To reach the optional effect of the different policies, member states have to be involved as well. It is important to make clear that there is no one-size-fits-all recipe. All the ingredients are being developed at the European level through the various relevant policies, but more work needs to be done to show how they can be used, combined, hacked, and improved to fit local tastes. There is in the coming years a large need for workshops, targeted communication campaigns, and info sessions to carry out this task. With its reach, the ENRD, or its future successor, could play a key role in this, but it would be necessary to go beyond the usual rural development actors and also bring in digital platforms, financial institutions, academia, and others.

Innovation

There is a wealth of relevant research being carried out under both national and EU-funded programs. Bioeconomy, new technologies, digital, artificial intelligence (AI), Blockchain, and many other emerging areas have the potential to be real game changers in rural communities.

But the knowledge needs to get out and work in the real economy. Good examples exist for bioeconomy, where big EU-funded demonstrators are making a real difference in the local economy, with entrepreneurial ecosystems developing around them. The same could become the case for the EU digital hubs; however, this remains to be seen.

There does seem to be a gap, however, when it comes to improving the use and take up of rurally relevant research. Whereas agri-food and forestry research benefits from the EIP-AGRI to link it to the world of practice — farmers and their advisors — the same is not the case for rural research in general. Here there could be a role for the innovation broker as proposed by the Economic and Social Committee but it could also be useful to link the future research program, Horizon Europe, better to the various existing networks and knowledge hubs for rural actors, in the same way as is done for the EIP-AGRI. This could be a task for the successor of the ENRD but it could also be useful to do a thorough screening of the networks and actors involved in rural innovation systems and decision-making today.

Rural Proofing

Rural proofing, as proposed by the Cork 2.0 Declaration, can be a relevant tool in EU policymaking. Specifically, it could help guide the European Semester process as well as the future programming of EU funds.

Yet, it is not only EU spending policies that affect rural areas. Rural proofing can also help develop specific schemes that can be conducive for smart villages and subsequently for rural well-being. This could for instance concern different corporate taxation rates for rural areas. Most EU directives and regulations affect rural areas (as well as all other parts of the Union). The same applies to national legislation. In some cases, rural and urban areas are impacted equally, yet sometimes there is a specific rural bias caused by geography, demography, or socioeconomic structures. This should be present in the mind of those legislating, and rural proofing can help with this.

References

Bled Declaration. (2018). *Bled Declaration for a smarter future of the rural areas in EU*. Retrieved from http://pametne-vasi.info/wp-content/uploads/2018/04/Bled-declaration-for-a-Smarter-Future-of-the-Rural-Areas-in-EU.pdf. Accessed on April 2018.

Committee of Regions. (2017a). *The need for a White Paper on Rurality from a local and regional perspective for a European Rural Agenda after 2020*. Retrieved from https://cor.europa.eu/Documents/Migrated/Events/Brochure%20White%20Paper%20on%20Rurality%20EN.pdf. Accessed on August 2018.

Committee of Regions. (2017b). *Revitalisation of rural areas through smart villages*. Retrieved from https://cor.europa.eu/en/our-work/Pages/OpinionTimeline.aspx?opId=CDR-3465-2017. Accessed on May 2018.

DBIS. (2013, October). *Smart cities,* Background paper, Department for Business, Innovation and Skills (DBIS), United Kingdom. Retrieved from https://assets.publishing.service.gov.uk/government/uploads/system/uploads/attachment_data/file/246019/bis-13-1209-smart-cities-background-paper-digital.pdf. Accessed on September 2018.

European Commission. (2011). *The role of regional policy in the future of Europe*. Retrieved from https://ec.europa.eu/regional_policy/sources/docgener/panorama/pdf/mag39/mag39_en.pdf. Accessed on September 2018.

European Commission. (2012). *European innovation and partnership 'agricultural productivity and sustainability'*. Retrieved from https://ec.europa.eu/eip/agriculture/en/european-innovation-partnership-agricultural. Accessed on September 2018.

European Commission. (2015a). *Europe 20202 strategy*. Retrieved from https://ec.europa.eu/digital-single-market/en/europe-2020-strategy. Accessed on May 2018.

European Commission. (2015b). *Better regulation "toolbox"*. Retrieved from http://ec.europa.eu/smart-regulation/guidelines/docs/br_toolbox_en.pdf. Accessed on October 2018.

European Commission. (2016). *CORK 2.0 Declaration 2016 for a better life in rural areas*. Retrieved from https://ec.europa.eu/agriculture/sites/agriculture/files/events/2016/rural-development/cork-declaration-2-0_en.pdf. Accessed on April 2018.

European Commission. (2017a). *Study on broadband coverage in Europe 2017*. Retrieved from https://ec.europa.eu/digital-single-market/en/news/study-broadband-coverage-europe-2017. Accessed on December 2018.

European Commission. (2017b). *EU action for smart villages*. Retrieved from https://ec.europa.eu/agriculture/sites/agriculture/files/rural-development-2014-2020/looking-ahead/rur-dev-small-villages_en.pdf. Accessed on April 2018.

European Commission. (2017c). *Smart eco-social villages for rural development*. Retrieved from https://ec.europa.eu/info/news/smart-eco-social-villages-rural-development-2017-apr-04_en. Accessed on April 2018.

European Commission. (2018a). *Future of the Common Agricultural Policy*. Retrieved from https://ec.europa.eu/info/food-farming-fisheries/key-policies/common-agricultural-policy/future-cap_en. Accessed on June 2018.

European Commission. (2018b). *A sustainable bioeconomy for Europe: Strengthening the connection between economy, society and the environment*. Retrieved from https://ec.europa.eu/research/bioeconomy/pdf/ec_bioeconomy_strategy_2018.pdf#view=fit&pagemode=none. Accessed on November 2018.

European Commission. (2018c). *Legal texts and factsheets on the EU budget for the future*. Retrieved from https://ec.europa.eu/commission/publications/factsheets-long-term-budget-proposals_en. Accessed on June 2018.

European Council on Foreign Relations (ECFR). (2016). *The revenge of the countryside*. Retrieved from https://www.ecfr.eu/article/commentary_the_revenge_of_the_countryside7156. Accessed on December 2018.

European Economic and Social Committee. (2017). *Villages and small towns as catalysts for rural development*. Retrieved from https://www.eesc.europa.eu/en/our-work/opinions-information-reports/opinions/villages-and-small-towns-catalysts-rural-development. Accessed on May 2018.

European Network for Rural development. (2017). *Smart villages portal*. Retrieved from https://enrd.ec.europa.eu/smart-and-competitive-rural-areas/smart-villages/smart-villages-portal_en. Accessed on April 2018.

European Parliament. (2018). *The Common Agricultural Policy in figures*. Retrieved from http://www.europarl.europa.eu/ftu/pdf/en/FTU_3.2.10.pdf. Accessed on November 2018.

Lindblom, C. E. (1959). The science of "muddling through". *Public Administration Review, 19*(2), 79–88.

Lytras, M. D., & Visvizi, A. (2018). Who uses smart city services and what to make of it: Toward interdisciplinary smart cities research. *Sustainability, 2018*(10), 1998. doi:10.3390/su10061998

Matthews, A. (2007). *Rural development in the European Union: Issues and objectives.* Paper presented at Workshop 1 Public Policy and Rural Development – An EU/US Comparison Penn State/ERS Project: Design and Evaluation of Public Policies For Rural Development, Withersdane Hall Conference Center at the Wye campus of Imperial College, London (Ashford, Kent, United Kingdom), June 25 and 26. Retrieved from http://citeseerx.ist.psu.edu/viewdoc/download?doi=10.1.1.629.1033&rep=rep1&type=pdf. Accessed on November 2018.

R. E. D. (2017). For a European Rural Agenda post 2020! Proposal, Rurality-Environment-Development (R. E. D). Retrieved from https://cor.europa.eu/Documents/Migrated/Events/For%20a%20European%20Rural%20Agenda%20post%202020.pdf. Accessed on May 2018.

Sedov, V. A., Chelyshkov, P., Rujitskaya, A., & Solntseva, M. (2016). 'Smart city' (the European concept of 'smart city'). *SSRN Electronic Journal.*

Sherry, E., & Shortall, S. (2018). Methodological fallacies and perceptions of rural disparity: How rural proofing addresses real versus abstract needs. *Journal of Rural Studies.*

Shortall, S., & Alston, M. (2016). To rural proof or not to rural proof: A comparative analysis. *Politics & Policy, 44,* 35–55.

TFEU, Art 174. (2008). *Consolidated version of the treaty on the functioning of the European Union – Part three: Union policies and internal actions – Title XVIII: Economic, Social and territorial cohesion – Article 174 (ex Article 158 TEC).* Retrieved from https://eur-lex.europa.eu/legal-content/EN/TXT/?uri=CELEX%3A12008E174. Accessed on June 2018.

TFEU. (2007). *Consolidated versions of the treaty on European Union and the treaty on the functioning of the European Union.* Retrieved from https://eur-lex.europa.eu/legal-content/EN/TXT/HTML/?uri=CELEX:12012E/TXT&from=EN. Accessed on May 2018.

Visvizi, A., & Lytras, M. D. (2018a). Rescaling and refocusing smart cities research: From mega cities to smart villages. *Journal of Science and Technology Policy Management, 9*(2), 134–145. doi:10.1108/JSTPM-02-2018-0020

Visvizi, A., & Lytras, M. D. (2018b). It's not a fad: Smart cities and smart villages research in European and global contexts. *Sustainability, 2018*(10), 2727. doi:10.3390/su10082727

Visvizi, A., Lytras, M. D., Damiani, E., & Mathkour, H. (2018). Editorial: Policy making for smart cities: Innovation and social inclusive economic growth for sustainability. *Journal of Science and Technology Policy Management, 2018*(9), 126–133.

Chapter 3

Smart Villages Revisited: Conceptual Background and New Challenges at the Local Level

Oskar Wolski and Marcin Wójcik

Introduction[1]

The social and economic significance of rural areas is best understood when their development is interpreted as a process based on local resources (Wolski, 2019). In this context, the key to understanding rural area development is the manner in which these resources are used, and at the same the unique nature of each given rural area. An examination of rural areas leads to the analysis of differences between rural areas and consequently the nature and extent of various challenges associated with development as well as political and planning intervention at many different levels of government. This yields a situation where the planning of rural development or creating policies that largely satisfy various groups of stakeholders living in different rural areas is difficult and requires flexible solutions (OECD, 2006).

The implementation of new development policies including ones based on strategies of innovativeness is related largely to the specificity of places and

---

[1]In the following parts of the chapter, 'Smart Villages in European Union Policy – Context and Definition' and 'Challenges and (Pre)conditions for Implementation at the Local Level,' alongside the references and one's own knowledge, information obtained during O. Wolski's participation in meetings of the Thematic Group on Smart Villages, acting as part of the European Network for Rural Development (ENRD) in Brussels, were used. Hence, these parts of the chapter may partly express the opinions of the participants of the meetings as well. O. Wolski's participation in these meetings was financed by the Foundation for Assistance Programmes for Agriculture FAPA and the Agricultural Advisory Centre in Brwinów, Branch Office in Warsaw, serving as the Central Unit of the National Rural Network (KSOW) in Poland. The chapter reflects the authors' view, which does not purport to reflect the opinions of KSOW.

resources. The idea of 'place' is well understood (Buttimer, 1976; Relph, 1976; Tuan, 2001), but attention is now focused primarily on the local scale, which is associated with the analysis of local social and economic relationships – both formal and informal. This relationship-based way of understanding spatial differences leads to the identification of the specificity of place via the identification of leading determinants of the occurrence of various social and economic processes (Rauch, 1993; Romer, 1990). Ideas related to research on the specificity of place underscore the problem of triggering actions that may be considered their specialization. Hence, the introduction of innovation ought to be closely linked with specialization strategy in order to create local competitive advantage (Bilbao-Osorio & Rodríguez-Pose, 2004; Fagerberg, Verspagen, & Caniéls, 1997). McCann and Ortega-Argilés (2015) employ in these cases concepts such as embeddedness, relatedness, and connectivity. These are defined in the analyzed context in the following manner:

- Embeddedness: substantial link between local residents and their local area, which in this case implies their active interest in the current state of their commune in the area of finance, social issues, and spatial order. New technologies including new channels of social communication should not replace existing channels but should help improve the level of communication including yielding better circulation of information.
- Relatedness: its role is based on the formation of a group of people and institutions in pursuit of predefined priorities in the area of solving local development problems. The functioning of this type of network relies on the exchange of information that may be used to optimize the process of establishing a local development strategy and enjoying the benefits of an improved quality of life, which is where the internet plays a key role.
- Connectivity: the key role in this regard is played by the readiness of both people and institutions to face challenges posed by geographic space. Connectivity is based on the availability of modern technologies – and adaptation to changing information transfer capabilities as well as changes in the development of virtual spaces. However, mobility is the product of the need to learn about and create new places, which yields a societal awareness of the existence of diversity in a world based on the rapid flows of people and goods (McCann & Ortega-Argilés, 2015).

All three concepts may be examined in relation to the proliferation of information and its employment in the creation of knowledge. Research has shown that the role of geographic space is important in the transfer of knowledge, which means that it occurs primarily at the local level – as local as the neighborhood level – and not at more aggregated, regional levels (Andersson, Klaesson, & Larsson, 2016; Koster, van Ommeren, & Rietveld, 2014). The core of this issue lies in increasing networking as an important characteristic of modern economic and social life, especially in the era of rapidly developing new technologies. On the one hand, it is the mobility of man across geographic space and the ease with which migration decisions are made, even when facing significant social issues and physical distance. On the other hand, we have virtual mobility or the capability to operate in a parallel reality (McCann & Ortega-Argilés, 2015).

Hence, social and technological changes frequently prompt policymakers to choose such a direction in the planning of rural development – a direction reliant not only on the endogenous approach, but one which also factors in the reality that no single policy can solve all problems (Naldi, Nilsson, Westlund, & Wixe, 2015; OECD, 2006; Ward & Brown, 2009). The desired approach is that of smart rural development, which is fundamentally based on the creation of appropriate conditions for development, as opposed to the detailed planning of development in itself (Naldi et al., 2015).

The implementation of the idea of smart development is based largely on the social and economic nature of a given geographic area – and especially on the nature of the commune as a local community. The unique nature of each given commune largely determines the various development goals of the commune and the very need for innovation in a given area. Attention is paid to the susceptibility of a place (i.e., rural commune) to the occurrence of a certain type of technological or social innovation and its adoption by the local community and institutions. In other words, an innovation will not be adopted if a given local community does not have the potential (need and skills) for its adoption. Innovations are useless if they are not new; however, they are also useless if they are 'so new' that they cannot be understood. The introduction of an innovation should therefore be closely linked with the social and economic specialization of the given commune. Also, the innovations will be different depending on the exact function of each given commune – whether a given commune is mostly agricultural or mostly non-agricultural or perhaps it is primarily a residential area or performs still some other social or economic function. Hence, the needs of local residents, institutions, and business entities in particular should determine the basis for formulating the principles of smart development.

In the context of the European Union (EU) policy, the smart development of rural areas is often discussed as part of the development of smart villages. The discussion of this approach to the rural development results from not only the earlier mentioned rural development conditions but also from efforts designed to target EU policy toward rural places and the needs of rural communities (ENRD, 2018).

Nevertheless, the smart village approach to development is associated clearly with the notion of innovativeness – which in itself may be treated as a challenge in rural areas – as well as a change in the mindset of stakeholders thinking about the implementation of solutions that may bring about an improvement in the quality of rural life (Wolski, 2018a) and a reevaluation of existing patterns of action. Although rural development is a process that involves different levels of government from local to global (OECD, 2006; van der Ploeg et al., 2000), *action always occurs at the local level*, at specific sites and via specific individuals. In this context, we describe changes in mindset and the reevaluation of existing strategies as key challenges in the implementation of smart development in the management of rural development at the local level. For the very same reason, it is necessary to assign a special role to local governments in smart rural development as a result of the following set of conditions.

First, local governments are delegated special powers to manage the process of development at the local level. Second, they are largely responsible for the acquisition of EU funding, which plays a major role in the development of European villages. Third, innovation-based, smart development becomes more relevant in the context of the formulation of EU policies (Wójcik, 2018), which means that local governments need to act in a manner that will allow them to function amidst changes in EU policy, and pursue benefits from the EU to the greatest extent possible. This perception of the role of local government is associated not only with its responsibility for rural development but also with a recognition of the attributes of local government (Wolski, 2018a).

The chapter has two aims. First is to try to answer the following question: *How can local governments – and to some extent other key actors – create the right conditions for the development of smart villages at the local level?* Second is *to discuss the aforesaid smart village approach in the context of selected theories of rural development.* Hence, we are also going to provide a discussion of selected theoretical issues associated with the notion of rural development and the evolution of these issues. In the third part of the chapter, we discuss smart villages as an approach to rural development. The basis for this discussion is a debate on smart development pursued by the EU, which attaches a practical element to the theoretical discussion. The fourth part of the chapter examines practice versus theory in terms of the actions that local governments and other actors may take in order to generate smart development. The concluding section includes a number of recommendations for policymakers.

Selected Concepts in Contemporary Rural Development

Smart villages as an approach to rural development represent an example and the effect of the evolution of a number of theoretical and practical approaches to development first and foremost at the local level and to some extent regional level. These range from exogenous approaches such as the agglomeration model to endogenous approaches such as the local milieu model (Dawkins, 2003; Terluin, 2003). Modern development is characterized by a high degree of complexity, rapid change, overlap between local and global factors, transfer of knowledge, and diffusion of innovation. Hence it appears that attempts at explanation based on endogenous approaches alone are insufficient, and an examination of the issue additionally demands a mixed and neo-endogenous approach – for example, territorial innovation models (Terluin, 2003).

It is not possible to present at least a majority of theories of development of rural areas, even in the form of a review, associated with all of the discussed approaches. In this part of the work, we will attempt to describe what we believe to be the most important theories in terms of smart villages. This attempt is also a type of generalization and does not exhaust the subject area, yet it does illustrate the theoretical basis of smart villages.

Endogenous Approach

The dominant approach to rural development since the late 1970s has been the endogenous approach – having replaced the exogenous approach (Terluin, 2003). The latter assumes that the development of rural areas is determined by external factors such as cities, which is why the development process may be structured independently of the local characteristics of a given rural area (Slee, 1994; Wolski, 2019). According to the exogenous approach (rural studies perspective), the development of rural areas may be anchored to agglomeration models of regional development (regional studies perspective) (Terluin, 2003). However, the endogenous approach focuses on local development rooted in local factors and local resources; therefore, endogenous development may be identified with local development (Morris, 1998; Picchi, 1994; Terluin, 2003).

According to Lowe, Murdoch, and Ward (1995), in the context of the formulation of rural area development policy, this translates into support for the diversification of the local business sector and the rural economy in general, efforts pursued via a grassroots approach, local community initiatives and other local endeavors, and appropriate training programs and opportunities for professional development. Given that the endogenous approach also respects local values (Slee, 1994), this may be linked with a 'revitalization paradigm' for rural areas. It may be considered a counterweight to the 'modernization paradigm,' which used to serve as the urban-industrial model for rural areas, where rural areas were examined through the lens of the social and cultural characteristics of cities, and development was supposed to occur through a transfer of urban-industrial schemes to rural areas (Kaleta, 1992; Wójcik, 2012).

The number of models of rural development that fit the endogenous approach is large.[2] For the purpose of this chapter, it is most important to discuss the concept of immobile resources and community-led rural development, as these concepts are closely linked with the notion of smart rural development (smart villages). First, these linkages are due to the fact that the 'smart' label is usually associated with innovativeness as well as research, knowledge, and education (Naldi et al., 2015), all of which lead to innovativeness and may be considered immobile resources (Bryden & Dawe, 1998). Second, people are really the key 'ingredient' in the smart village approach (EC, 2017a; ENRD, 2018) – or essentially local governments defined as the community governing at this level (Wójcik, 2018) and other stakeholders working together.

---

[2]In addition to those already discussed in the chapter are sustainable development, eco-development, community-led local development and LEADER, rural renewal, multifunctional development, modernization, as well as models that are related to social capital, urban-rural partnerships, social participation, economics of culture, and governance.

Third, these concepts fit the general model of local milieu development (Terluin, 2003; Wójcik, 2016). Regional policy is thought currently to be more effective in stimulating growth than sector policy and this includes growth in rural areas (Ward & Brown, 2009), while EU policy as a whole is evolving in the direction of territorial policy. Thus, the conceptual linkage between regional and rural development appears to be relevant in order to shift the academic debate toward specialized recommendations for policymakers, stakeholders, and other actors.

According to Bryden and Dawe, a rural area's competitive advantage comes from key resources that remain locally rooted despite the onset of globalization and easy access to mobility. These immobile resources are said to include four types of capital: social, cultural, environmental, knowledge (1998). However, it would be difficult to argue that success in rural area development depends solely on these types of resources. Hence, it may be inferred that a linkage between these resources and mobile resources as well as the proper utilization of both types of resources also play an important role. Mobile resources include labor, financial capital, information, and other transferable goods and services (Bryden & Dawe, 1998; Terluin, 2003). In other words, while resources are important in development when they are defined as the total potential of a geographic area including its material goods, an area's competitive advantage is built upon its immobile resources, which are not subject to competition. Hence, a given area's possession of certain immobile resources, as opposed to their acquisition through transfer or migration, yields a competitive advantage virtually immediately.

It is important to note that community-led rural development is not exactly the same as community-led local development (CLLD), which has become commonplace thanks to EU funding in the form of Rural Development Programme (RDP), although both development paths imply the involvement of local communities in the development process. Community-led rural development first and foremost refers to the skills, knowledge, and potential of a community (McArdle, 2012; Murray & Dunn, 1996), where it coincides with community development (cf. Green & Haines, 2012). These community attributes allow it to create potential for self-organization and mutual assistance manifested in the use of specialized knowledge in the course of processes managed by groups, conflict resolution skills, leadership, understanding the interests of other local actors, and achieving a common vision (Murray & Dunn, 1996).

However, the creation of community potential is not an independent process and needs the support of appropriate institutions at various levels and the establishment of partnerships especially at the local level as well as the readiness of institutions managing development to join grassroots initiatives (Dej, Janas, & Wolski, 2014; Terluin, 2003). Local governments can not only be a part of community and establish local networks but may also actively participate in the creation of the potential of the entire rural community by employing its attributes.

CLLD based on LEADER method principles is largely a 'practical' idea, as it allows local communities to suggest initiatives, participate in both the preparation and completion of projects, and help make development decisions associated with dedicated financial instruments and tools. It is a mechanism used by the EU (EC, 2017b).

Mixed Exogenous/Endogenous Approach and Neo-endogenous Approach

The mixed approach to rural area development, which includes the neo-endogenous approach (Wolski, 2019), shows a close linkage with changes occurring across rural areas today as well as with their unique nature resulting from the EU's common development policy. Terluin (2003) identified three types of forces affecting modern rural areas. The first is territorial dynamics unique to a given geographic area that include regional issues, structures, entrepreneurial traditions, public and private networks, work ethics, regional identity, community participation level, and attractiveness of the cultural and natural environment. The second force is population dynamics including natural growth and migrations. Rural communities are home to native and in-migrant populations, with the second group deserving particular attention. This latter group has been shown to impact the local economy, especially its regeneration (Bosworth, 2006; Herslund, 2011; Stockdale, 2005, 2006). In addition, in-migrants represent a diverse pool of people whose lifestyles often differ from those of the local population and interactions with them may pose a challenge to local cultural norms (Terluin, 2003; Wójcik, 2016). In effect, the management of the development process must factor in the diversity of rural actors and their various goals (Flyn & Marsden, 1995).

While the first two forces apply to local and regional growth patterns, the third dynamics are global in nature, although it is superimposed on key local and regional patterns to some extent (Terluin, 2003). Undoubtedly rural areas are affected by globalization, as are urban areas, and this process connects the local and the global (Wolski, 2019). Hence, the success of rural areas in the development process is increasingly related to their relationship with the non-local or a host of social and economic changes at the country level and the world level (Ward & Brown, 2009).

The mixed exogenous/endogenous approach suggests a joint impact of local and non-local development factors by placing rural development in the context of globalization driven by a host of technological changes including means of communication and information processing techniques increasingly employed in the service of economic goals (Terluin, 2003). Hence, the neo-endogenous approach may be treated as a product of the mixed approach, as it implies that local development may be animated from three possible directions, separately or together: (1) within the local area via grassroots work or local actors, (2) from above by national or EU-based factors, and (3) from the intermediate level particularly in the form of nongovernmental organizations (Ray, 2001). Other key elements of this approach include cultural assets, uniqueness of place, and ability of the local population to introduce desired changes (Bosworth & Atterton, 2012; Ray, 2000; Ward et al., 2005).

Most Important Reasons for the Evolution of Rural Development Policies

Rural areas characterized by significant diversity starting at the local or regional level leading up to the continental level or higher experience rapid changes, which is no longer a debatable issue. The debate is rather focused on the nature of each constituent change including economic transformation and social transformation driven by a variety of development factors including global issues as well as the change in the societal perception of rural areas.

The rural economy is increasingly less reliant on agriculture (Baldock, Dwyer, Lowe, Petersen, & Ward, 2001; Galdeano-Gómez, Aznar-Sánchez, & Pérez-Mesa, 2011; Marsden, 1999; van der Ploeg et al., 2000). This is also reflected in rural development policies that promote the emergence of non-agricultural sectors designed to support rural multifunctionality. Baldock et al. (2001) showed that: (1) local actors represent a wide array of interests in line with their social identification and economic status, which is why the interests of farmers are not always the same as the interests of their parent rural areas, (2) the multifunctionality of rural areas is dependent on their internal diversification and external pressures – and is not simply the result of the structure of the agricultural sector, (3) the competitiveness of the agricultural sector does not always imply an economic vitality of rural areas, as vitality is not completely reliant on the economic and social activity of farmers. This notion of rural areas is strongly reflected in the "New Rural Paradigm" (OECD, 2006), which suggests that the basic goal of development work should be to increase the competitiveness of rural areas, assess the quality of their assets, and activate unused resources. Given the complexity of the process, local and regional government entities need to be involved, as should be national governments and international organizations. The aforementioned also implies an important distinction – what is needed to accelerate rural growth is investment and not subsidies (OECD, 2006). However, changes in the rural social fabric may be associated with urbanization and counterurbanization (Boyle & Halfacree, 1998). In later stages of development, changes are based on global processes such as consumerism, individualism, and also changing styles of communication.

These constituents of rural change affect the way rural areas are perceived. Today rural areas are no longer perceived as material production spaces or one-dimensional, simple, and easily understandable spaces. Rural areas are also perceived as consumer spaces (Cloke, 2006) that include tourism, recreation, cultural and social activity, and environmental value (Crouch, 2006; Halfacree, 2012; Marsden, 1999; Murdoch & Pratt, 1993; Wójcik, 2016).

Given the aforementioned issues, researchers are focusing on the dynamics and direction of change in rural areas, which are related to each other to a greater or lesser degree. The key question is not how to fit into a changing reality, but how to: (1) face challenges resulting from the dynamics of change in rural areas, their relationship with cities, their institutional environment, and the social and economic conditions in which they function, and (2) respond to the direction of change with the protection of good and resource which the countryside is in mind (Wolski, 2018a). At the same time, these changes are not taking place in the same manner in all rural areas, as these are highly diversified. Hence, rural areas not only are different from one another but also respond differently to change, which may be explained by the susceptibility to certain social and economic processes described in the first part of the chapter. By utilizing their variable characteristics, different rural areas develop via a number of different development mechanisms, and their unique characteristics may be said to be determinants of their development (Wolski, 2019).

In summary, the evolution of Rural Development Policy is based on the diversity of rural areas, dynamics of change, and directions of change. The currently emerging response to the first issue is that one size does not fit all (OECD, 2006; Ward & Brown, 2009). This new mindset is reflective of the conviction that no single policy can address the challenges of all rural areas and consequently policies need to be targeted toward local characteristics (Naldi et al., 2015). There is a growing awareness in rural area development research that changes not only yield risks but also opportunities (ENRD, 2018). This is why development policy should not only address risk management but first and foremost should focus on exploiting the full potential of rural areas, which would also increase their ability to grow and prosper. This is also why the current thinking is that the stimulation, management, and planning of rural development needs to be based on characteristics such as innovativeness, flexibility, embeddedness, and potential for catalyzing positive future change as well as economic efficiency (Wolski, 2018a).

Smart Villages in European Union Policy – Context and Definition

The real world including the rural context is not easy to generalize and generalizations are not useful in most cases in the formulation of solutions for specific geographic areas. This requires out-of-the-box thinking about rural development and its planning. At the same time, it would be harmful to operate completely outside of the political and economic context, as solutions produced in this way would not be practical and could not be implemented on a broader scale. Policy is designed to serve as a problem resolution tool. It is not meant to create new problems. Hence, rural development efforts discussed in the previous paragraphs may be reduced to two key concepts: innovativeness and pragmatism (Wolski, 2018a). In other words, new solutions need to be built on existing foundations, as policy formulation is a continuous process and it is characterized by evolution, as is the development process itself.

During the 2014–2020 programming period, the European Network for Rural Development (ENRD) has been working, among others, on the 'Smart and Competitive Rural Areas' theme. It responds to the three EU rural development policy priorities:

(1) fostering knowledge transfer and innovation in agriculture, forestry, and rural areas;
(2) enhancing farm viability and competitiveness of all types of agriculture in all regions and promoting innovative farm technologies and the sustainable management of forests; and

(3) promoting food chain organization, including processing and marketing of agricultural products, animal welfare, and risk management in agriculture.

Work on these issues includes specific subthemes: Food and Drink Supply Chains, Rural Businesses, and Smart Villages.

Smart villages relate mostly to the first of the mentioned EU policy priorities, where innovativeness in rural areas is core. This brings together 'smart' and 'innovative.' So, what is the nature of the innovativeness of the idea of smart villages? First, smart villages may be understood as communities that refuse to wait for change and its benefits – and instead choose to take the initiative (ENRD, 2018). This is why their innovativeness relies upon the shaping of attitudes from reactive to proactive. This type of attitude also represents a view opposite to that of entitlement, where the key beneficiaries of support programs are merely passive recipients and not stakeholders or actors. This proactive attitude is the cornerstone of the idea of smart villages. As part of smart development, rural communities explore a variety of solutions to the problems they are facing and attempt to find opportunities for development (ENRD, 2018). It is such villages that may be able to handle the dynamics of change and direction of change in modern rural areas. Different villages may be faced with different challenges – what is important is how they respond to these challenges (Wolski, 2018a).

Given the fact that the idea of smart villages, which may include computerization and innovativeness, is applicable first and foremost to the way entities respond to change and challenges, and not to specific areas of action, the definition of this term must not be too narrow (cf. Visvizi & Lytras, 2018a). This is due to the fact that any attempts at classification are case-dependent (Zavratnik, Kos, & Stojmenova Duh, 2018). In addition, it is the context of action that must be factored in for every case of 'smart' action (Visvizi & Lytras, 2018b). This approach underscores the diversity of rural areas and the common nature of their problems. The formulation of a definition that could be adapted to different contexts of action served as one of the tasks of a special thematic group (TG) established at the ENRD in the years 2017–2019 at the contact point in Brussels – the TG on Smart Villages. This group meets on a regular basis to discuss changes in EU policy for the development of European rural areas, examine opportunities to support smart villages as part of various financing programs, exchange experiences and best practices in smart rural development, and provide a platform for dialog between the representatives of various groups of stakeholders (Wolski, 2018a).

The definition of smart villages is focused on certain characteristics and attitudes of communities (cf. Zavratnik et al., 2018). To be a 'smart village' means the following (ENRD, 2018, p. 7):

- use of digital technologies whenever these help in attaining established goals; at the same time, these technologies are not the only tools used in the process;
- thinking outside of the rural context; while all initiatives are rooted at the local level, smart communities do understand that rural areas function as part

of some surrounding area including linkages with other rural areas and cities; it is these linkages that need to be activated;
- establishing new forms of cooperation and new networks of stakeholders that may include farmers and other rural actors, local governments, the private sector, and the community as a whole; the work of such alliances is facilitated by both grassroots work and top-down management; and
- independent thinking; there does not exist a one standardized model of smart villages or a closed set of tools used by smart villages. This is why smart villages need to assess their resources, use the best available knowledge, and take the initiative in shaping their local reality.

Work on the definition of smart villages began in the context of the characteristics and attitudes of communities described earlier as part of a pilot program dedicated to smart villages called the *Pilot Project on Smart Eco-Social Villages*. The discussion focused initially on the definition proposed by the European Commission (2017a, p. 3), whereby smart villages are "rural areas and communities which build on their existing strengths and assets as well as on developing new opportunities" where "traditional and new networks and services are enhanced by means of digital, telecommunication technologies, innovations and the better use of knowledge." The members of the TG as well as other interested parties had the opportunity to comment and make suggestions as to how to help improve this definition. As of October 2018, work on the definition is ongoing.[3]

While the definition of smart villages is not yet complete, two basic conclusions may be drawn at this point in time. First, the definition remains open in order to factor in the great diversity of rural areas found across Europe and their different needs. This approach will most likely be maintained in the future. At the same time, this definition creates a certain framework for analysis and action, which is critical for the practical functioning of smart villages and from the perspective of rural area planning. Second, the term 'smart' does not need to imply high-tech projects, or the projects that concentrate on innovative, infrastructural developments, or the project that addresses only the needs specific to advanced rural areas (Wolski, 2018a). This approach may also be adapted to the introduction of services in rural areas as well as job creation and the meeting of basic social needs. This is the pragmatic aspect of this view of rural development – while innovative, many key challenges at the local level affecting smart villages (communities) are 'typical' and affect many rural areas and communities that continue to retain their own local uniqueness. Nevertheless, these problems remain unsolved, and this is why existing ways of thinking about development need to be abandoned in favor of a reevaluation of development strategies.

---

[3]Progress in this area may be tracked on the website of the project: http://pilotproject-smartvillages.eu (accessed on October 2018). The website also provides basic information on the subject of smart villages.

The pragmatism of this approach is also manifested in available tools as well as financing instruments. The TG on Smart Villages recommends that any existing tools and instruments be utilized, especially those part of the RDP. This may include the following measures: M19 – support for the LEADER local development (CLLD), M07 – basic services and village renewal in rural areas, M16 – cooperation, and also parts of M6 – farm and business development. The TG also notes the role of local action groups that often initiate action at the local level, and in a way approach village residents with a palette of initiatives. Village residents play a central role in the smart village approach.[4]

In summary, smart villages is an approach to rural development that is strongly rooted in contemporary theories of rural development and is the result of current thinking associated with EU policy. This approach is consistent with ideas discussed earlier: (1) the rejection of the one-size-fits-all mentality, (2) the activation of currently underused endogenous potentials of rural areas with consideration of their nature and diversity, and (3) the evolution of a mutual responsibility mindset based on the involvement of institutions at various levels and the engagement of various groups of stakeholders.

Challenges and (Pre)conditions for Implementation at the Local Level

In the era of financial support for rural areas from the EU, it appears to be most important to establish structures, environments, and climates at the local level in order to help implement the model of the smart village. Local governments play a key role in this process, as they have the needed authority and obligations to the local community (Wolski, 2018a). However, other stakeholders may also actively participate. Research has shown that in order to achieve results at the EU level via effective policy, it is necessary to generate a wide array of competencies at the local level (Barca, 2009). This assertion provides a basis for analysis in the next part of the chapter.

As mentioned early on in the chapter, the key challenges at the local level are the change in the attitudes of actors with respect to rural development and the reevaluation of existing ways of doing things. The first challenge implies certain schools of thought that drive relations between development stakeholders. The second challenge implies that local communities need to use certain resources in order to create the right conditions for the implementation of smart development.

How Should Smart Villages Think about Local Development?

The smart village approach leads to a number of innovative solutions that are associated with certain risks, which are also inherent in the rural transformation stage

---

[4]Additional information on financial support and other forms of support for smart villages is found in Wolski (2018a).

Europe is currently in. Hence, the creation of the right conditions for the functioning of smart villages demands not only the acceptance of a certain level of risk but also a shift in thinking about both success and failure in project completion (Wolski, 2018a). Effective projects that are valuable in terms of how they serve the needs of local communities often require modifications. The method of trial and error serves local communities well in terms of how they learn – and requires that they be ready to change. Change is also expected of development policy including that at the EU level, which needs to provide appropriately flexible solutions. This is why this challenge concerns not only the local level, which is our main subject of interest, but also an entire cross section of development management levels.

Innovation should not encounter resistance in rural areas only due to the fact that it is safer to realize standard projects (Idziak & Wilczyński, 2003; Zajda, 2015) or projects that can be fully planned and whose results are easy to predict. The established mentality is due to a number of factors that include a rigid financing system – a commonly cited reason – and the administrative and legal relationship and informal relationship between local governments and rural communities as well as representatives of nongovernmental organizations and business entities (Wolski, 2018a). The involvement and cooperation of all stakeholders should yield cocreation of solutions and not the establishment of lobbies and groups of private interests. Hence, at the local level where everyone knows everyone, it is important to act in the interest of everyone involved and not merely pursue private interests.

Moreover, the meeting of obligations by local governments in relation to local residents triggers innovativeness and creativity, and builds mutual trust. When basic needs of residents are not being met, they become priorities for residents, which is quite natural. However, in this type of case, local residents are not able to think about their needs in a non-standard manner, and this includes the utilization of external funding (Wolski, 2018b).

In this context, a smart village is one where the local government meets its obligations to its constituents, which is why they are able to express their own, less-standardized ideas and engage in creative interactions whose purpose is to realize these ideas, which are supported by the local government via its active participation using a variety of communication channels – including participation based on information and communications technology (ICT) solutions and real participation (Wolski, 2018a). When local communities are not limited and are free to express themselves, they follow novel development pathways that also expand into the public sector and business sector. This leads to an exchange of ideas that generates a proactive linkage, which helps build competitive advantage in places where it is seemingly absent. In effect, what used to constitute a need of the local community becomes a catalyst for change and a generator of new ideas. Yet, this type of interaction is not possible in the absence of mutual trust. The degree to which the triggering of innovativeness in rural projects is possible is not large, as shown by the small percentage of such projects.[5]

---

[5]Based on studies conducted in Poland (Wolski, 2018b).

How Should Smart Villages Make Use of Resources?

The number of determinants of smart development is large and includes a wide array of local infrastructural issues, social capital, economic potential, culture, and the natural environment (Naldi et al., 2015; Nooteboom, 2001). If the determinants are to be treated as local resources, then how should they be utilized in order to build competitive advantage in the development process (cf. Bryden & Dawe, 1998)? Hence, it appears that at the local level it is necessary to be aware of how these resources can be used. This part of the chapter discusses resources that may be increasingly common in rural areas, although may be varied at the local level, and may require a new way of managing them.

The first relevant factor is technical infrastructure including easy access to broadband internet and also places where it can be used. In addition to the more obvious roles played by the internet in smart development, it can be used to reduce the communication distance that separates rural area residents and local governments as well as to access sources of knowledge (Wójcik, 2018). This particular goal can be facilitated by: (1) clarity and functionality of local government web pages including the option to use them to communicate with local residents – that is, dialog with local government officials, forums for key groups of stakeholders, and access to information on currently relevant social issues, (2) adaptation of public consultation to an electronic format, as an additional channel, which may include the posting of surveys and the opportunity to speak out online, all of which is designed to reduce conflict in real life, (3) expansion of public e-services, which would also improve opportunities for a realization of various goals online – that is, conducting official business online, both cultural and educational aims, social assistance goals, community management issues, and other issues, and (4) creation of expanded versions of geoportals that allow users to collect information on spatial management and local investment directions.

Smart development is also impacted by access to institutions associated with knowledge such as schools and other educational facilities (Ward et al., 2005). Research has shown that a rich cultural offering also helps facilitate smart development – and this includes local culture (Bell & Jayne, 2010). These are additional areas where local communities can take action on the road to smart development and these efforts do not necessarily involve significant financial expenditures. The modern role of the school need not be limited to its mandatory educational mission, but may serve as an instrument for general community education. Local governments should make it a priority to create the right conditions for the expansion of competencies and acquisition of new knowledge via organization of training sessions and informational meetings for representatives of various businesses as well as publication of brochures and booklets.

Likewise, the triggering of activity including that in the realm of culture and not in any direct manner related to social and economic development also helps facilitate smart development, as it stimulates reserves of creativity and helps develop this characteristic over the long term. Hence, local cultural events may be used as places to exchange ideas, network with various groups of stakeholders, advertise local products and services, and get to know local community leaders, all of which ultimately results in the transfer of local knowledge or knowledge that is relevant to other members of the local community. Local events that strengthen traditions and local identity will also help strengthen the roots of local actors and enhance local products and services that are unique in one way or another. The resources and ways of utilizing them in the smart development of rural areas discussed in this section of the chapter are related to contexts noted by McCann and Ortega-Argiles (2015) who described smart specializations such as connectivity, embeddedness, and relatedness, which may confirm how these affect the development of rural areas.

Local governments and other representatives of the local community have influence in terms of how modern infrastructure is used, how knowledge capital is created, and how culture can be used for networking purposes, wherein all the while culture in itself is helped to flourish. At the same time, innovativeness at the local level is also facilitated. Most of the aforementioned constitute key resources needed to create competitive advantage. As smart villages represent an approach to rural development focused on the challenges of a given community living in a geographic area, actors in the local community can use their proactive stance to combat challenges by relating to rural localities (Halfacree, 2006; Hoggart, 1990).

Conclusions and Recommendations

Smart villages are an approach to rural development whose central role is played by the local community and its actions. Rural communities consist of local governments and other rural actors representing various interests. The role played by the representatives of this community does not determine their level of involvement, as this approach is based on the concept of cocreation. Actions taken are designed to match the needs of the community (human-centered approach), as this approach is designed to counter key challenges associated with rural life in their most realistic form. Finally, realization on the local scale helps match projects with local conditions – a place-based approach.

Smart villages are smart communities capable of responding to challenges associated with change in rural areas. As the very nature of the challenge is not important, but the response is important, this is also an approach to rural development that may be adapted independently of the characteristics of the given rural area. This corresponds to contemporary notions of rural development part of the broadly defined neo-endogenous school of thought as well as with the direction of the evolution of rural area policy and politics in general, which stress that development is the responsibility of many different entities and levels of government.

As noted earlier in the chapter, a village developing according to the principles of smart development functions at the local level in an institutional, social, and cultural environment in which certain characteristics are prominent, and are determined by local governments as well as the local community. On the basis of the aforementioned theoretical concepts and selected challenges, we wish to provide a number of recommendations for the implementation of the smart village approach at the local level:

- *Smart rural development practiced at the local level must constitute a means of thinking about rural area development in general.* This requires not only changes in the perception of innovation at the local level, as discussed earlier, but also the emergence of this approach in local development strategies on a practical level. In other words, it is necessary to identify a certain direction in the school of thought on rural development and strengthen its legal basis, which will serve as the foundation for further action.
- Local governments constitute the local community – as defined by the principle of cocreation – and possess predefined competencies and attributes invaluable in the implementation of smart development. Yet, *the democratization of action at this level should manifest itself in the reduction of the distance in the relationship between local governments and other representatives of the local community who are treated as stakeholders.*
- Local governments that meet the basic needs of the local community and consider the needs of various stakeholders not only fulfill their mandated mission but also build trust and help release the creative potential of the local community. In this scenario, the community is freed from thinking only about current needs, which is a path to the emergence of new, bold ideas. However, *thinking about bold ideas needs to be preceded by an evaluation of current actions and identification of real needs.*
- The implementation of smart development must not ignore seemingly less relevant efforts that are fundamental in nature, as they relate to the proper utilization of key resources already available locally. *Smart villages seek and find opportunities resulting from their own local potential: properly utilize information and communications technology, and put best practices to work. While they remember about their local identity, they develop horizontal relationships (work with other villages) and vertical relationships (work with other institutions managing development at other levels).*

All of the aforementioned allows for the formation of attitudes and characteristics of the local community, which have served as the starting point for the definition of smart villages/communities. This full circle of analysis not only concerns the issues raised in this chapter but also describes the nature of the implementation of the smart village approach, which emerges from each given local community and is implemented in the name of the same community. *Being a smart village first and foremost implies local government efforts and efforts of cooperating institutions that lead to a greater transparency of decisions made by local governments elected by local communities. Second, it implies an increase in the quality of life as a result of innovations designed for each particular rural area.*

References

Andersson, M., Klaesson, J., & Larsson, J. P. (2016). How local are spatial density externalities? Neighbourhood effects in agglomeration economies. *Regional Studies, 50*(6), 1082–1095.

Baldock, D., Dwyer, J., Lowe, P., Petersen, J. E., & Ward, N. (2001). *The nature of rural development: Towards a sustainable integrated rural policy in Europe: A ten-nation scoping study for WWF and the GB countryside agencies (Countryside Agency, Countryside Council for Wales, English Nature and Scottish Natural Heritage)*. London: Institute for European Environmental Policy.

Barca, F. (2009). *An agenda for a reformed cohesion policy: A place-based approach to meeting European Union challenges and expectations*. Retrieved from www.europarl.europa.eu/meetdocs/2009_2014/documents/regi/dv/barca_report. Accessed on October 2018.

Bell, D., & Jayne, M. (2010). The creative countryside: policy and practice in the UK rural cultural economy. *Journal of Rural Studies, 26*(3), 209–218.

Bilbao-Osorio, B., & Rodríguez-Pose, A. (2004). From R&D to innovation and economic growth in the EU. *Growth and Change, 35*(4), 434–455.

Bosworth, G. (2006). Counterurbanisation and job creation: Entrepreneurial in-migration and rural economic development. *Centre for Rural Economy Discussion Paper Series, 4*, 1–15.

Bosworth, G., & Atterton, J. (2012). Entrepreneurial in-migraton and neoendogenous rural development. *Rural Sociology, 77*(2), 254–279.

Boyle, P., & Halfacree, K. (Eds.). (1998). *Migration into rural areas: Theories and issues*. Chichester: Wiley.

Bryden, J., & Dawe, S. (1998). *Development strategies for remote rural regions: What do we know so far?* Retrieved from http://nibio.academia.edu/johnbryden. Accessed on November 2017.

Buttimer, A. (1976). Grasping the dynamism of lifeworld. *Annals of the Association of American Geographers, 66*, 277–292.

Cloke, P. (2006). Conceptualizing rurality. In P. Cloke, T. Marsden, & P. Mooney (Eds.), *Handbook of rural studies* (1st ed., pp. 18–28). London: Sage.

Crouch, D. (2006). Tourism, consumption and rurality. In P. Cloke, T. Marsden, & P. Mooney (Eds.), *Handbook of rural studies* (1st ed., pp. 355–364). London: Sage.

Dawkins, C. J. (2003). Regional development theory: Conceptual foundations, classic works, and recent developments. *Journal of Planning Literature, 18*(2), 131–172.

Dej, M., Janas, K., & Wolski, O. (Eds.). (2014). *Towards urban-rural partnerships in Poland: Preconditions and potential*. Kraków: Institute of Urban Development.

European Commission. (2017a). *EU action for smart villages*. Retrieved from https://ec.europa.eu/agriculture/sites/agriculture/files/rural-development-2014-2020/looking-ahead/rur-dev-small-villages_en.pdf. Accessed on September 2018.

European Commission. (2017b). *Guidelines: Evaluation of LEADER/CLLD*. Brussels. Retrieved from https://enrd.ec.europa.eu/file/11452/download_en?token=CwgJOvg5. Accessed on October 2018.

European Network for Rural Development. (2018). Smart villages: Revitalising rural services. *EU Rural Review 26*. Luxembourg: Publications Office of the European Union. Retrieved from https://enrd.ec.europa.eu/sites/enrd/files/enrd_publications/publi-enrd-rr-26-2018-en.pdf. Accessed on September 2018.

Fagerberg, J., Verspagen, B., & Caniéls, M. (1997). Technology, growth and unemployment across European regions. *Regional Studies, 31*(5), 457–466.
Flyn, A., & Marsden, T. (1995). Guest editorial. *Environment and Planning, A, 27*, 1180–1192.
Galdeano-Gómez, E., Aznar-Sánchez, J. A., & Pérez-Mesa, J. C. (2011). The complexity of theories on rural development in Europe: An analysis of the paradigmatic case of Almería (south-east Spain). *Sociologia Ruralis, 51*(1), 54–78.
Green, G. P., & Haines, A. (2012). *Asset building and community development* (2nd ed.). Los Angeles, CA: Sage.
Halfacree, K. (2006). Rural space: Constructing a three fold architecture. In P. Cloke, T. Marsden, & P. Mooney (Eds.), *Handbook of rural studies* (1st ed., pp. 44–62). London: Sage.
Halfacree, K. (2012). Diverse ruralities in the 21st century: From effacement to (re-)invention. In L. J. Kulcsar & K. J. Curtis (Eds.), *International handbook of rural demography* (1st ed., pp. 387–400). Dordrecht: Springer.
Herslund, L. (2011). The rural creative class: Counterurbanisation and entrepreneurship in the Danish countryside. *Sociologia Ruralis, 52*(2), 235–255.
Hoggart, K. (1990). Let's do away with rural. *Journal of Rural Studies, 6*(3), 245–257.
Idziak, W., & Wilczyński, R. (2003). *Odnowa wsi: Przestrzeń, ludzie, działania* [Rural renewal: Space, people, action]. Warszawa: Fundacja Programów Pomocy dla Rolnictwa FAPA.
Kaleta, A. (1992). Assumptions of renewal of the rural areas of Europe. In M. Wieruszewska (Ed.), *Renewal of the rural areas: Between myth and hope* (pp. 13–39). Warsaw: Institute of Rural and Agriculture Development of Polish Academy of Science.
Koster, H. R. A., van Ommeren, J., & Rietveld, P. (2014). Is the sky the limit? High-rise buildings and office rents. *Journal of Economic Geography, 14*(1), 125–153.
Lowe, P., Murdoch, J., & Ward, N. (1995). Networks in rural development beyond exogenous and endogenous models. In J. D. van der Ploeg & G. van Dijk (Eds.), *Beyond modernisation: The impact of endogenous rural development* (pp. 87–105). Assen: Van Gorcum.
Marsden, T. (1999). Rural futures: The consumption countryside and its regulation. *Sociologia Ruralis, 39*(4), 501–520.
McArdle, K. (2012). What makes a successful rural regeneration partnership? The views of successful partners and the importance of ethos for the community development professional. *Community Development, 43*(3), 333–345.
McCann, P., & Ortega-Argilés, R. (2015). Smart specialization, regional growth and applications to European Union cohesion policy. *Regional Studies, 49*(8), 1291–1302.
Morris, A. (1998). *Geography and development* (1st ed.). London: UCL Press.
Murdoch, J., & Pratt, A. (1993). Rural studies: Modernism, postmodernism and the "post rural". *Journal of Rural Studies, 9*(4), 411–427.
Murray, M., & Dunn, L. (1996). Capacity building for rural development in the united states. *Journal of Rural Studies, 1*(1), 89–97.
Naldi, L., Nilsson, P., Westlund, H., & Wixe, S. (2015). What is smart rural development? *Journal of Rural Studies, 40*, 90–101.

Nooteboom, B. (2001). *Learning and innovation in organizations and economies* (1st ed.). Oxford: Oxford University Press.
OECD. (2006). *The new rural paradigm: Policies and governance.* Paris: OECD.
Picchi, A. (1994). The relations between central and local powers as context for endogenous development. In J. D. van der Ploeg & A. Long (Eds.), *Born from within: Practice and perspectives of endogenous rural development* (pp. 195–203). Assen: Van Gorcum.
Rauch, J. E. (1993). Productivity Gains from Geographic Concentration of Human Capital: Evidence from the Cities. *Journal of Urban Economics, 34*(3), 380–400.
Ray, Ch. (2000). Endogenous socio-economic development in the European Union – Issues of evaluation. *Journal of Rural Studies, 16*(4), 447–458.
Ray, Ch. (2001). *Culture economies: A perspective on local rural development in Europe.* Centre for Rural Economy. Retrieved from https://www.ncl.ac.uk/media/wwwnclacuk/centreforruraleconomy/files/culture-economy.pdf. Accessed on October 2018.
Relph, E. (1976). *Place and placelessness* (1st ed.). London: Pion.
Romer, P. M. (1990). Endogenous technological change. *Journal of Political Economy, 98*(5), 71–102.
Slee, B. (1994). Theoretical aspects of the study of endogenous development. In J. D. van der Ploeg & A. Long (Eds.), *Born from within: Practice and perspectives of endogenous rural development* (pp. 184–194). Assen: Van Gorcum.
Stockdale, A. (2005). Incomers: Offering economic potential in rural England. *Journal of the Royal Agricultural Society of England, 166*, 1–5.
Stockdale, A. (2006). Migration: Pre-requisite for rural economic regeneration? *Journal of Rural Studies, 22*, 354–366.
Terluin, I.J. (2003). Differences in economic development in rural regions of advanced countries: an overview and critical analysis of theories. *Journal of Rural Studies, 19*, 327–344.
Tuan, Y.-F. (2001). *Place and space: The perspective of experience* (8th ed.). Minneapolis, MN: University of Minnesota Press.
van der Ploeg, J. D., Renting, H., Brunori, G., Knickel, K., Mannion, J., Marsden, T., ... Ventura, F. (2000). Rural development: From practices and policies towards theory. *Sociologia Ruralis, 40*(4), 391–408.
Visvizi, A., & Lytras, M. D. (2018a). Rescaling and refocusing smart cities research: From mega cities to smart villages. *Journal of Science and Technology Policy Management (JSTPM).* doi:10.1108/JSTPM-02-2018-0020
Visvizi, A., & Lytras, M. D. (2018b). It's not a fad: Smart cities and smart villages research in European and global contexts. *Sustainability, 10*(8), 2727.
Ward, N., Atterton, J., Kim, T.-Y., Lowe, P., Phillipson, J., & Thompson, N. (2005). Universities, the knowledge economy and 'neo-endogenous rural development'. *Centre for Rural Economy Discussion Paper Series, 1*, 1–15.
Ward, N., & Brown, D. (2009). Placing the rural in regional development. *Regional Studies, 43*(10), 1237–1244.
Wójcik, M. (2012). *Geografia wsi w Polsce: Studium zmiany podstaw teoretyczno-metodologicznych* [Rural geography in Poland: Study of changes in theoretical-methodological foundations]. Łódź: Wydawnictwo Uniwersytetu Łódzkiego.
Wójcik, M. (2016). Selected problems of contemporary socio-spatial changes in peri-urban areas of the city of Łódź (Poland). *Geographia Polonica, 89*(2), 169–186.

Wójcik, M. (Ed.). (2018). Inteligentny rozwój obszarów wiejskich (smart rural development): koncepcja, wymiary, metody [*Smart rural development: Concept, aspects, methods*]. Łódź: Global Point.

Wolski, O. (2018a). Smart villages in the EU policy: How to match innovativeness and pragmatism? *Village and Agriculture*, *4*(181), 163–179.

Wolski, O. (2018b). Problem of (non-) innovativeness in village renewal projects. *Acta Universitas Lodziensis, Folia Geographica Socio-Oeconomica*, *31*, 17–37.

Wolski, O. (2019). The place of rural areas in regional development concepts and processes. In P. Nijkamp & K. Kourtit (Eds.), *Rurality in an urbanized world*. Maastricht: Shaker.

Zajda, K. (2015). Rural areas as a territory for Innovation. *Village and Agriculture*, *3*(168), 7–19.

Zavratnik, V., Kos, A., & Stojmenova Duh, E. (2018). Smart villages: Comprehensive review of initiatives and practices. *Sustainability*, *10*(8), 2559.

Chapter 4

Toward a New Sustainable Development Model for Smart Villages

Raquel Pérez-delHoyo and Higinio Mora

Introduction

The emergence of the new scenario proposed by the 2030 Agenda for Sustainable Development (UN, 2015) — approved in 2015 by the United Nations General Assembly — invites us to reflect on the sustainability of the current development model and the need to create a new one through the achievement of the Sustainable Development Goals (SDG).

The SDG include "Goal 11: Make Cities and Human Settlements Inclusive, Safe, Resilient and Sustainable." Thus, 2030 Agenda recognizes that not only sustainable management and development of the urban environment are essential for the quality of life but also for human settlements in rural areas and villages.

To support the achievement of this Goal, Information, and Communication Technology (ICT) involved in the smart villages concept — action launched by the European Commission in 2017 — has much to offer (Zavratnik, Kos, & Duh, 2018). The expansion of ICT also offers great potential for improving the quality of life of communities in rural areas (Akca, Sayili, & Esengun, 2007; Mora, Gilart-Iglesias, Pérez-delHoyo, & Andújar-Montoya, 2017). However, in recent years, ICT implementation has been lower in rural areas (Szeles, 2018). The digital divide between the urban and rural environment has decreased considerably for some technologies such as mobile phones, but it is still maintained for other basic technologies, such as connection to asymmetric digital subscriber line (ADSL) lines.

According to data from the European Union Statistics Office — Eurostat — in 2014, 40.2% of people in the European Union (EU) were living in cities — densely-populated areas; 27.8% were in rural areas — thinly-populated areas; and 32.0% were in towns and suburbs — intermediate density areas (Eurostat, 2017). This means that rural environments are home to more than half of the EU's population and cover more than three-quarters of its territory. For 50 years, there have been different trends between rural and urban areas:

- Most urban regions continue to experience population growth, while the number of people living in many peripheral and rural regions is declining.
- Cities have a 'pull effect' mainly due to the perception of increased educational and employment opportunities.

Migration to cities from rural areas is becoming increasingly important, but rural areas are essential for achieving the EU's objectives of "smart, sustainable and inclusive growth" (European Commission, 2010). Urban and rural areas enjoy different but complementary assets, and the interrelations among them are important for economic viability and sustainable development (Gutman, 2007; Ward & Brown, 2009). The territorial dimension of EU legislation is therefore one of the political priorities. In this sense, the European Committee of the Regions intends to contribute to reducing the knowledge gap between regions and cities as a means of reducing the gap between urban and rural (Eurostat, 2017). In order to achieve this objective, ICT plays a fundamental role, introducing the concept of smart villages.

Rural society is increasingly open to a globalized world, and all the opportunities offered by this new world scenario must be taken advantage of (Dammers & Keiner, 2006; Westhoek, Van den Berg, & Bakkes, 2006). Many rural areas are experiencing processes of demographic reactivation, successfully competing in this environment, through the diversification of economic activity in accordance with new social demands and trends. However, other rural areas are facing this situation in a decline marked by an aging population, depopulation, poor access to a range of services, etc. In all these cases, the technological gap with the urban areas is evident.

There are important differences between the rural and urban territories of the EU member states, so there is no one-size-fits-all solution. Strategies must arise from the definition of the characteristics and potentials of each area or grouping of areas (Badulescu & Badulescu, 2017; Davidova & Bailey, 2014; Pašakarnis, Morley, & Malienė, 2013; Prestia & Scavone, 2018). For this reason, in-depth research is needed to get to know the nature and needs of each area and, consequently, to design an effective smart strategy. This research is not possible without the active participation of rural areas, with the main problems arising in rural areas with fewer resources.

A smart strategy aims to attract communities and investors that may be interested in the unique potential of rural areas to develop their ideas and projects. However, to achieve this purpose, it is necessary to have a deep knowledge of rural areas, their history and values that identify them and differentiate them from other areas, in other words, from the rural potential that positions them in a significant place in the territory.

In order to streamline the debate, this chapter proposes to develop a methodology based on resilience. Resilience is defined as the ability of a habitat or system to recover to its initial state when the disturbance to which it has been subjected has ceased. All rural areas that are currently in decline have had a better past and can renew themselves by building on the link to their better past (Antrop, 2005). In this regard, a retrospective of rural areas is proposed based on the experience of the 'garden city' model, for which the advantages of rural areas were evident over those of urban areas (Lenzi & Perucca, 2018). The aim is to reconsider the intrinsic and differential qualities of rural areas in order to recover and enhance them with the added value of the EU Smart Villages approach. These facets will be drivers of sustainable development.

The remaining part of this chapter is organized as follows. Section 'Toward a Smart Village Model: A Reality in the Twenty-first Century' briefly describes the proposals that — in the context of the EU's rural development policies — have led to the concept of EU smart villages and their definition. 'Smart Strategies for Smart Villages' introduces the idea of smart strategy. 'A Methodology for the Design of a Smart Strategy' proposes a methodology for the design of a smart strategy: the principles and theoretical framework on which it is based, as well as basic guidelines for villages with fewer resources to begin designing their smart strategy. Finally, the last section outlines conclusions and recommendations.

Toward a Smart Village Model: A Reality in the Twenty-first Century

The key question for rural development on how to make living and working in rural areas attractive was addressed in the Cork Declaration in 1996. The Cork Declaration recognized that rural areas and their inhabitants are a real asset for the EU because they have the capacity to be competitive. Its potential is based on its own nature characterized by a unique cultural, economic, and social fabric, as well as a wide variety of activities and landscapes (Cork Declaration, 1996). In particular, the Cork Declaration pointed to the need to reverse rural migration and to respond to the increasing demands in all areas of life: personal development, health, well-being, security, and leisure. In short, it denounced the imbalance in investment in infrastructure, as well as in education, health, and communication services, between rural and urban areas.

Twenty years later, when the next generation of politicians and representatives meet again in Cork, the first Declaration of 1996 becomes even more relevant. The exodus of young people and depopulation aggravate the situation in rural areas. There is a greater need for policies to ensure that villages and rural areas remain attractive places to live and work, by improving access to services and opportunities and by encouraging traditional and new-sector entrepreneurship. The Cork 2.0 Declaration (2016) declared the 10 policy orientations which should guide the EU's innovative, integrated, and inclusive rural and agricultural policy to take advantage of the potential of rural areas: promoting rural prosperity, strengthening rural value chains, investing in rural viability and vitality, preserving the rural environment, managing natural resources, encouraging climate action, boosting knowledge and innovation, enhancing rural governance, advancing policy delivery and simplification, and improving performance and accountability (Cork 2.0 Declaration, 2016). This future vision of rural areas introduces a new component, the use of ICT. The focus is not only on overcoming the digital divide between rural and urban areas but also, and especially, on developing the potential of connectivity and digitization of rural areas.

Cork 2.0 Action Plan (2017) introduces the concept of smart villages. The EU Action for Smart Villages is thus launched, announcing a series of initiatives in the fields of rural development, regional development, research, transport, energy, policies, and digital services. EU Action for Smart Villages defines the concept of intelligent villages as referring to rural areas and communities that build on their existing strengths and assets, as well as on the development of new opportunities. Thus, in intelligent villages, traditional and new networks and services are enhanced through ICT and better use of knowledge for the benefit of people and businesses. The concept of intelligent villages does not propose a single solution. It is based on the needs and potential of each territory and is driven by a strategy (European Commission, 2017).

Recently, the Bled Declaration (2018) has given a definitive boost to the use of digital technologies to address the challenges facing rural areas in Europe. In short, the smart village initiative can be the model of the twenty-first century for making rural areas places where people can and want to live because digital innovations make their lives easier and more comfortable – an ongoing issue since Cork's first declaration in 1996.

Smart Strategies for Smart Villages

Although advances in ICT have led to a rich body of research on smart cities, in the context of smart villages the research developed is still incipient. That is why initiatives are needed to introduce the idea of smart villages into the scientific debate. In this sense, a potent discussion on possible smart strategies for smart villages has recently been developed in a multidisciplinary way (Visvizi & Lytras, 2018). This chapter aims to add to this initiative.

According to the definition provided by the European Commission, smart villages refer to rural areas that find the keys to their development in their specific context. Through dialog with the environment, they find a unique profile that sets them apart from other areas and positions them in the context of the global economy. Thus, a smart village knows which functions it performs better than any other rural village.

The future of rural areas depends on the identification of their vocation in relation to their environment. They must know the identity signs and the components of excellence that define their own smart profile, from which they can design their smart strategy – without forgetting the opportunities that can arise from integrating into natural spaces, as well as rural and urban systems of their environment.

All the declarations and action plans referred to in the previous section, as well as in the European Commission's own definition of smart villages, converge on the same fundamental idea that the potential of rural areas is based, above all else, on the value of their own rural identity: their culture, their people, and their economy. Therefore, knowing and recovering this identity is the first step toward enabling rural areas to design smart strategies that will enable them to become unique focal points of attraction.

In this sense, this chapter provides a methodology for rural areas to design their smart strategies. It is a simple methodology that can be used to ensure that rural areas with fewer resources are not left behind and can also start designing their own smart strategy from scratch at no excessive cost. It consists of an elementary theoretical framework and basic but effective guidelines or tools.

The design of a smart strategy requires a process of reflection in which the distinguishing components specific to each village or group of villages emerge. These components of excellence will serve to provide growth perspectives and will be the driving force behind the villages of the future.

The objective is that villages with fewer resources also have a reference point for how to approach their own reflection, not in an isolated way but within the common framework of work that this simple methodology provides. The background to this methodology is therefore a strategic approach to promoting the development of as many villages as possible with the smart village approach.

A Methodology for the Design of a Smart Strategy

The proposed methodology is based on the principle of resilience. Resilience is defined as the ability of a habitat or system to recover to its initial state when the disturbance to which it has been subjected has ceased. Based on the fact that all rural areas that are currently in decline have had a better past, this methodology proposes to build on the link with the best past in order to begin the reflection that will enable rural areas to design their own smart strategy. In order to design the strategy, it is therefore necessary to have an in-depth knowledge of rural areas, their history and potential, identifying the values that differentiate them from other areas and positions them in the territory. Recognizing the rural potential, as well as the main weaknesses and needs, is therefore the first step.

Only on the basis of this knowledge can urgent actions be defined. Actions that, on the one hand, strengthen the components of excellence of the different rural areas and, on the other hand, improve the deficiencies and needs identified. These actions, which can be fundamentally addressed with the massive use of ICT, benefit the recovery of the potential of rural areas but at the same time open up opportunities for communities that can turn these actions into their future projects – in a given sector of activity or services (Knickel et al., 2018; Pölling et al., 2017; Rivza & Kruzmetra, 2017).

The signs of identity and the components of excellence of each rural area will be the focal points of attraction for all types of communities and investors that may be interested in the unique potential of rural areas to develop their ideas and projects. However, a thorough prior dissemination of information on both the potential and the areas of opportunity for action in each of the villages or groups of villages in a territory is essential. In this sense, the development of a common catalog, structured by regions and territories, of rural areas and their potentials and opportunities, for the development of smart villages projects of different nature – of landscape, healthy, productive, and heritage interest – is an ambitious project with great potential.

Therefore, the methodology consists of three phases, as shown in Figure 1:

(1) recognize the rural potential, as well as the weaknesses and development needs of rural areas;
(2) definition of actions to strengthen the components of excellence and reduce the weaknesses of rural areas; and
(3) dissemination of information on smart villages – development of a catalog of potentials and opportunities.

The different actors involved in the process are at least the following:

- The active participation of the smart villages communities is needed.
- Collaboration is also needed with research groups that provide techniques and methodologies for research and dissemination of results.
- Support is also needed from the relevant political and governance structures to provide funding and process verification mechanisms.
- Finally, the initiative of other communities that find in the smart villages the opportunity to develop their future projects in areas that can be as varied as education, services, and the environment is needed.

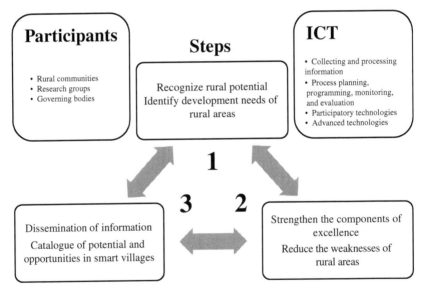

Figure 1. A Methodology for the Design of a Smart Strategy. *Source*: Authors' own figure.

In each of the phases the use of ICT is fundamental, distinguishing among others the following technologies:

- technologies that benefit the active participation of smart village communities such as mobile applications, web applications, social networks, blogs, and wikis;
- technologies for data collection and information processing;
- technologies for planning, programming, monitoring, and assessment of processes; and
- advanced technologies specific to each specific process or smart village project developed.

Theoretical Framework

The proposed methodology is based on a first decisive step: the recognition of the rural potential, as well as the weaknesses and main needs of rural areas. This first step can be addressed by rural areas with fewer resources, to the extent of their means, and is the beginning of the design of their smart strategy.

It consists of a retrospective action based on the theoretical framework of the garden city model, for which the advantages of rural areas were evident over those of urban areas, becoming in themselves points of attraction. The relationship between the countryside and the city has always been a recurrent debate.

The garden city model was conceived by Ebenezer Howard in the late nineteenth century as a solution to the problems of the industrial city (Howard, 1902). Of course, the problems of the twenty-first century are different, but the model and its methodology are still of relevance today (Friedman, 2018).

In this proposal, the garden city is of interest as a concept. It is a model that competed directly with the urban model, from which two aspects are fundamentally highlighted:

(1) It develops a territorial model, where rural areas play an important role. All of these areas are organized on a territorial basis and are connected by efficient transport infrastructures.
(2) Its main objective is social innovation, where there is room for a new organization of communities, distribution of responsibilities, formulas of self-management, self-government, new business models, and funding.

The garden city model defines a new mode of habitat, which seeks the union of the goodness of the countryside and the city. This concept was explained by Howard through his "three magnets" theory (Figure 2). In a simple diagram it reflected the pros and cons of living in the country or the city. However, a third habitat, the garden city, was defined as a third magnet that attracted people by bringing together the advantages of living both in the city and in the countryside.

Obviously, this theory was utopian, but reconsidered from the present time, from the perspective of the technological era of the twenty-first century and the possibilities offered by the connectivity and digitalization of rural areas, the perspective

Town   **Country**

Closing out of nature–Social opportunity
Isolation of crowds–Places of amusement
Distance from work–High money wages
High rents and prices–Chances of employment
Excessive hours–Army of unemployed
Fogs and droughts–Costly drainage
Foul air and murky sky–Well-lit streets
Slums and gin palaces–Palatial edifices–Crowded dwelling

Lack of society–Beauty of nature
Hands out of work–Land lying idle
Trespassers beware–Wood Meadow forest
Long-hours-low-wages–Fresh air–low rents
Lack of drainage–Abundance of water
Lack of amusement–Bright sunshine
No public spirit–Need for reform
Deserted villages

The people
Where will they go?
Town–Country

Social opportunity–Beauty of nature
Fields and parks of easy access
Low rents–High wages
Low rates–Plenty to do
Low prices–No sweating
Field for enterprise–Flow of capital
Pure air and water–Good drainage
Bright homes and gardens–No smoke–No slums
Freedom–Co-operation

Figure 2. The 'Three Magnets' Theory – Garden City Model. *Source*: Authors' own figure, on the basis of *Garden Cities of Tomorrow* (Howard, 1902).

is different. If we also take into account current social trends (Cork Declaration, 2016), it is worth reflecting today on this model conceived 120 years ago.

- People are increasingly paying attention to quality of life and rural areas are in a unique position to respond to these interests and provide the basis for a genuine modern and quality development model.
- The concept of public support for rural development, harmonized with the proper management of natural resources and the maintenance and enhancement of biodiversity and cultural landscapes, is gaining increasing acceptance.

Smart villages can bring together the advantages of the countryside and the city, as many of the advantages of the city can now be incorporated into rural areas through ICT. With a massive ICT implementation effort it is possible to position smart villages as those places that bring together the advantages of rural and urban areas and, therefore, as the garden city model of the twenty-first century. This argument can be a point on which strategies to attract communities and activate the economy are based.

First Step in Defining a Smart Strategy

Starting from the premise that a smart village must be designed by the community itself, not by the markets, it is necessary that active communities capable of organizing themselves to reach a consensus on future projects live in the smart villages. In this sense, smart villages need the participation of the population as a civil society, as well as strong leadership to activate processes that guarantee innovation. Coherent political leadership is a competitive advantage for smart villages, but its lack can be filled by civil society, by the community itself, through foundations or associations, and also by the business community. Therefore, it is essential to attract such communities, for which smart villages must design their own smart strategy based on dialogue with the environment. The attractiveness of rural areas lies in their very nature and rural potential. However, today, thanks to ICT, the benefits traditionally offered by cities are also added (Figure 3).

The first step in defining a smart strategy is therefore to reconsider the intrinsic and differential qualities of rural areas in order to recover and enhance them with the added value of the EU Smart Villages approach. These facets will be focal points and drivers of sustainable development.

This retrospective action is sufficiently simple, so that it can be developed by rural areas that, in spite of their low resources, want to begin the design of their smart strategy. The following issues are proposed for in-depth reflection and basic guidelines are provided for their implementation:

- Environmental sensitivity and responsibility
  Environmental quality affects the quality of life and is a factor of uniqueness and competitiveness for the development and attraction of economic activities.

  Smart villages, as part of the territorial model, can take on a new, more ambitious ethic with regard to the natural environment. They are particularly sensitive to the problems and opportunities of the physical environment.
It is proposed to reflect on the following aspects and the essential enabling technologies:
  – the singularities of the physical environment;
  – its capacity to accommodate;
  – the vocation of its different areas; and
  – the value of planning, urban and housing design, and landscape architecture.

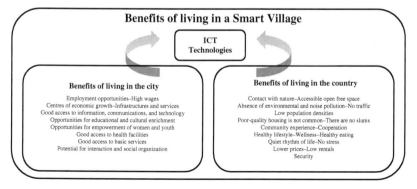

Figure 3. Benefits of Living in a Smart Village. *Source*: Authors' own figure.

- Commitment to social cohesion and development
  The renovation of the image of rural areas and public spaces of relationship and coexistence has obvious effects on the competitive and attractive capacity for the development of economic activities.
  Smart villages, made up of small communities, can work toward social balance and full inclusiveness, transparency, interaction, and cooperation.
  It is proposed to reflect on the following aspects and the essential enabling technologies:
  - actions that promote the democratization of decision-making processes and foster a sense of belonging and identity that fosters collective participation; and
  - active participation processes as mechanisms against exclusion and local development that foster a tolerant environment.

- Coherent structure of government and territory
  Smart villages, in proportion to the scale of their small communities, can have an efficient political and administrative structure.
  It is proposed to reflect on the following aspects and the essential enabling technologies:
  - the creation of appropriate bodies to carry out projects with specific objectives;
  - the possibilities offered by new technologies for managing basic services, speeding up administrative processes, providing services to citizens, fostering a sense of citizenship, and encouraging citizen participation; and
  - in particular, the possibilities offered by new technologies for access to health and education services, identifying the areas in which specific training is required.

- Work and innovation
  The most efficient way to create competitive advantage is to innovate. The key to territorial innovation is people. Smart villages that have learned from their own experience and that of other villages, and know their uniqueness and potential, offer interesting opportunities and can attract human capital.
  It is proposed to reflect on the following aspects and the essential enabling technologies:
  - educational needs and areas where specific training is needed as a prelude to innovation; and
  - possibilities for disseminating the unique profile and rural potential of smart villages, to attract people and workers with different levels of training and companies based on quality of life, a tolerant and innovative environment, and the existence of an interesting project for the future.

- Connections with other rural areas and cities
  In a global world, networks and flows are imperative. Smart villages can take advantage of all the opportunities offered by the globalized world thanks to the massive use of technology.

It is proposed to reflect on the following aspects and the essential enabling technologies:
- the active search for smart links and strategic alliances;
- participation in networks of rural areas and cities for common objectives, cultural harmony, size or geographical location, etc., which provide a strategic position – for example in the fields of healthy lifestyle, wellness, healthy eating, etc.; and
- the political, social, economic, cultural, and idea exchange that takes place at different levels – local, regional, national, and international.

Conclusions and Recommendations

In the context of competition and cooperation between cities and territories of the global economy, smart villages are able to create competitive advantage for the various activities – economic, residence, leisure, culture, social relations, etc. – thanks to technology.

People are the main resource for creating competitive advantage. For this reason, smart villages need active communities capable of organizing themselves to reach a consensus on future projects. Developing the capacity to educate and attract human resources is the most important challenge in smart villages.

Smart villages become attractive because of what they are, that is, because of their uniqueness – the unique and unrepeatable – and, in addition, because they have agreed on an smart project or strategy for the future. Therefore, all rural areas, without leaving any behind, should undertake the design of their own smart strategy to aspire to become a smart village.

This chapter provides a simple methodology and basic guidelines for villages with fewer resources to begin designing their smart strategy. This first step consists of a retrospective action, based on resilience – the recovery of the best past – that seeks to rediscover the rural potential, identifying the intrinsic and differential qualities of rural areas. The proposed methodology is inspired by the model of the garden city, for which the advantages of rural areas were evident over those of urban areas. The ultimate aim is to enhance these qualities with the added value of the EU Smart Villages approach.

In conclusion, the following five recommendations for the formulation of policies and/or action plans are presented:

(1) Developing the capacity to educate and attract human resources is the most important challenge in smart villages.
(2) All rural areas, without leaving any behind, should undertake the design of their own smart strategy and agree on a project for the future, within the common framework of the initiatives proposed by the EU.
(3) The first step in the design of a smart strategy consists of a retrospective action that seeks to rediscover the rural potential, identifying the intrinsic and differential qualities that position each of the rural areas or groups of

them in the territory. It is urgent to implement this first step, which can be implemented with not excessive resources.

(4) Support from policy and governance structures is needed to provide funding and verification mechanisms for the following circular process: (1) recognize rural potential and identify development needs of rural areas, (2) strengthen the components of excellence and reduce the weaknesses of rural areas, and (3) dissemination of information on smart villages.

(5) Development of a catalog, structured by regions and territories, of rural areas and their different potentials and opportunities, for the development of smart villages projects of different nature − of landscape, healthy, productive, heritage interest, etc. This may be an ambitious project with great potential.

References

Akca, H., Sayili, M., & Esengun, K. (2007). Challenge of rural people to reduce digital divide in the globalized world: Theory and practice. *Government Information Quarterly*, *24*(2), 404−413. doi:10.1016/j.giq.2006.04.012

Antrop, M. (2005). Why landscapes of the past are important for the future. *Landscape and urban planning*, *70*(1−2), 21−34. doi:10.1016/j.landurbplan.2003.10.002

Badulescu, D., & Badulescu, A. (2017). Rural tourism development through cross-border cooperation. The case of Romanian-Hungarian cross-border area. *Eastern European Countryside*, *23*(1), 191−208. doi:10.1515/eec-2017-0009

Bled Declaration. (2018). For a smarter future of the rural areas in EU, having regard to the conclusions of the meeting at Bled, Slovenia, on 13 April 2018, and previous declarations, such as the Cork 2.0 Declaration. Retrieved from http://pametne-vasi.info/wp-content/uploads/2018/04/Bled-declaration-for-a-Smarter-Future-of-the-Rural-Areas-in-EU.pdf

Cork 2.0 Action Plan. (2017). European Commission. Retrieved from https://ec.europa.eu/agriculture/sites/agriculture/files/events/2016/rural-development/cork-action-plan_en.pdf

Cork 2.0 Declaration. (2016). A better life in rural areas. Cork 2.0 European Conference on Rural Development. Retrieved from https://ec.europa.eu/agriculture/sites/agriculture/files/events/2016/rural-development/cork-declaration-2-0_en.pdf

Cork Declaration. (1996). A living countryside. The European Conference on Rural Development. Retrieved from http://www.terport.hu/webfm_send/545

Dammers, E., & Keiner, M. (2006). Rural development in Europe: Trends, challenges and prospects for the future. *disP-The Planning Review*, *42*(166), 5−15. doi:10.1080/02513625.2006.10556958

Davidova, S., & Bailey, A. (2014). Roles of small and semi-subsistence sarms in the EU. *EuroChoices*, *13*(1), 10−14. doi:10.1111/1746-692X.12044

European Commission. (2010). Europe 2020: A strategy for smart, sustainable and inclusive growth. Retrieved from http://eur-lex.europa.eu/legal-content/EN/TXT/PDF/?uri=CELEX:52010DC2020&rid=1

European Commission. (2017). EU action for smart villages. Retrieved from https://ec.europa.eu/agriculture/sites/agriculture/files/rural-development-2014-2020/looking-ahead/rur-dev-small-villages_en.pdf

Eurostat. (2017). Eurostat regional yearbook – 2017 edition. Publications Office of the European Union. Retrieved from https://ec.europa.eu/eurostat/documents/3217494/8222062/KS-HA-17-001-EN-N.pdf/eaebe7fa-0c80-45af-ab41-0f806c433763

Friedman, A. (2018). *Fundamentals of sustainable urban renewal in small and midsized towns*. Cham: Springer International Publishing AG.

Gutman, P. (2007). Ecosystem services: Foundations for a new rural–urban compact. *Ecological Economics, 62*(3–4), 383–387. doi:10.1016/j.ecolecon.2007.02.027

Howard, E. (1902). *Garden cities of tomorrow* (original 1898 title: Tomorrow: A peaceful path to real reform). London: Swan Sonnenschein & Co., Ltd.

Knickel, K., Redman, M., Darnhofer, I., Ashkenazy, A., Chebach, T. C., Šūmane, S., ... Rogge, E. (2018). Between aspirations and reality: Making farming, food systems and rural areas more resilient, sustainable and equitable. *Journal of Rural Studies, 59*, 197–210. doi:10.1016/j.jrurstud.2017.04.012

Lenzi, C., & Perucca, G. (2018). Are urbanized areas source of life satisfaction? Evidence from EU regions. *Papers in Regional Science, 97*, S105–S122. doi:10.1111/pirs.12232

Mora, H., Gilart-Iglesias, V., Pérez-delHoyo, R., & Andújar-Montoya, M. (2017). A comprehensive system for monitoring urban accessibility in smart cities. *Sensors, 17*(8), 1834. doi:10.3390/s17081834

Pašakarnis, G., Morley, D., & Malienė, V. (2013). Rural development and challenges establishing sustainable land use in Eastern European countries. *Land Use Policy, 30*(1), 703–710. doi:10.1016/j.landusepol.2012.05.011

Pölling, B., Prados, M. J., Torquati, B. M., Giacchè, G., Recasens, X., Paffarini, C., ... Lorleberg, W. (2017). Business models in urban farming: A comparative analysis of case studies from Spain, Italy and Germany. *Moravian Geographical Reports, 25*(3), 166–180. doi:10.1515/mgr-2017-0015

Prestia, G., & Scavone, V. (2018). Enhancing the endogenous potential of agricultural landscapes: Strategies and projects for a inland rural region of Sicily. In *Smart and sustainable planning for cities and regions: Results of SSPCR 2017* (Vol. 2, pp. 635–648). Cham: Springer International Publishing AG.

Rivza, B., & Kruzmetra, M. (2017). Through economic growth to the viability of rural space. *Entrepreneurship and Sustainability Issues, 5*(2), 283–296. doi:10.9770/jesi.2017.5.2(9)

Szeles, M. R. (2018). New insights from a multilevel approach to the regional digital divide in the European Union. *Telecommunications Policy, 42*(6), 452–463. doi:10.1016/j.telpol.2018.03.007

UN. (2015). Transforming our world: the 2030 Agenda for Sustainable Development. United Nations. Retrieved from https://sustainabledevelopment.un.org/content/documents/21252030%20Agenda%20for%20Sustainable%20Development%20web.pdf

Visvizi, A., & Lytras, M. (2018). It's not a fad: Smart cities and smart villages research in European and global contexts. *Sustainability, 10*(8), 2727. doi:10.3390/su10082727

Ward, N., & Brown, D. L. (2009). Placing the rural in regional development. *Regional studies, 43*(10), 1237–1244. doi:10.1080/00343400903234696

Westhoek, H. J., Van den Berg, M., & Bakkes, J. A. (2006). Scenario development to explore the future of Europe's rural areas. *Agriculture, ecosystems & environment, 114*(1), 7–20. doi:10.1016/j.agee.2005.11.005

Zavratnik, V., Kos, A., & Duh, E. S. (2018). Smart villages: Comprehensive review of initiatives and practices. *Sustainability, 10*(7), 2559. doi:10.3390/su10072559

Chapter 5

The Role of LEADER in Smart Villages: An Opportunity to Reconnect with Rural Communities

Enrique Nieto and Pedro Brosei

Introduction

The smart villages concept has the potential to allow communities from rural areas in the European Union (EU) to reap the wealth of opportunities available and to face contemporary challenges by putting people at the center of the development process. The EU Action for Smart Villages describes this concept as:

> rural areas and communities which build on their existing strengths and assets as well as new opportunities to develop added value and where traditional and new networks are enhanced by means of digital communications technologies, innovations and the better use of knowledge for the benefit of inhabitants.

Local people and innovation – both social and digital – are at the core of this newly developed concept in Europe (European Commission, 2017). The smart villages concept has gathered a lot of political interest geared toward reverting these negative tendencies. In other words, momentum now exists to support the vibrant development of rural areas across the EU by reconnecting with their communities.

There are key challenges in rural areas related to increasing discontent triggered by the growing disparities with cities in terms of quality jobs, the provision of basic services, depopulation, and increased vulnerability to external shocks.

Attempts to address these have been numerous, and LEADER[1] has contributed to ameliorate them to a great extent.

LEADER has been a key policy tool for local development in the rural areas of the EU since the 1990s. Its aim is to implement integrated rural development. Its application was meanwhile extended to other types of areas (coastal and urban) and is now part of the broader Community-led Local Development (CLLD) approach. From its inception, LEADER was understood as a radical shift from the earlier agricultural-centered policy paradigm toward the integrated development paradigm intended to bring new vitality to rural areas based on bottom-up initiatives (European Commission, 2006). It could be argued that aiming at 'smartness' is inherent in the LEADER approach since one of its key principles is to foster innovation, as well as to support community-led actions based on local integrated strategies.

While the smart villages concept and the LEADER approach share common key features, some small differences exist which create the space to complement each other in their practical dimensions. LEADER is a long-standing methodology which is embedded in the implementation system of European Agricultural Fund for Rural Development (EAFRD) Rural Development Programmes (RDPs) in member states, while smart villages appear as a new concept for development with a focus on smart solutions to support the long-term transition toward a sustainable future in rural areas. This leaves a lot of leeway to apply the smart villages concept through different policy measures and tools, out of which LEADER is probably one of the main ones. At this stage, the main challenge seems to be to clarify the links between smart villages and LEADER and understand how the latter can help rural communities become smart villages. To provide an answer, certain considerations should be explored.

Despite the fact that LEADER has always been a test field and incubator for innovative experiences with innovation at its heart, there are several reasons that explain why the links between the LEADER approach and the 'smartness' concept might not be self-evident:

- LEADER is not designed in most rural development programmes in a way that allows Local Action Groups (LAGs) to tap into innovation and more demanding local initiatives. In many cases, the regulatory framework designed in the member states restricts LEADER only to the implementation of standard measures. This hampers the development and support of collective, small and risky projects. LAGs are not given sufficient resources and capacity for deeper animation and facilitation of innovative processes, often due to the heavy administrative burden imposed by the regulatory framework (ENRD, 2018a). This is often linked to the 'consequences of mainstreaming' LEADER in the second pillar of the common agricultural policy (CAP).

---

[1]LEADER (Acronym derived from French: *Liaison Entre Actions de Développement de l'Économie Rurale*) is an acronym that stands for "links between actions for the development of the rural economy."

- Socioeconomic needs in some rural areas are very basic and outstanding, and these call for short-term solutions. In this context, LAGs do not prioritize thinking on 'smart' or on activities with a long-term focus on transition to a more sustainable future. In addition, the current common monitoring and evaluation system (CMES) for the EAFRD does not help, because it offers job creation as the only result indicator for LEADER (ENRD, 2018a). This indicator is driving policy actions supported under LEADER in many rural areas.

Despite the aforementioned constraints faced by LEADER, excellent examples exist of how LEADER currently applies the smart villages concept. These provide good arguments to use this long-standing approach as one of the key tools to implement smart villages in the next programming period in many member states. It will be the member states in their CAP plans who will decide their own way of using LEADER and smart villages in a coherent way.

Considering that outstanding challenges exist in rural areas, it is critical to build understanding and consensus among rural stakeholders – including academics – about the links between smart villages and LEADER. This is key considering the strong political will that exists to make smart villages a reality in the near future (European Commission and European Parliament, 2018).

This chapter aims to compare LEADER and smart villages so as to give more clarity about the interconnections between the two. To this end, the first section presents an analysis of the complementary features between the smart villages concept and the LEADER approach (and vice versa) in the field of integrated development policies. The next section of this chapter identifies the common boundaries between smart villages and LEADER by exploring existing LEADER initiatives in smart villages in areas such as energy, mobility, entrepreneurship, training, and broadband across Europe, and it outlines the various roles LEADER could play in the implementation of smart villages. Lastly, the chapter presents conclusions and recommendations with a particular aim to support policymakers in designing an appropriate policy framework that maximizes the added value brought by these indispensable concepts for local development.

Placing LEADER and Smart Villages in the Modern Rural Development Policy Paradigm

Smart villages and LEADER are both in some ways embedded in the international and European policy framework. Starting from the Sustainable Development Goals (SDGs) – as the highest-level framework for all development policies – the smart concept is incorporated across most of the policy areas covered by the SDGs such as well-being (inclusive and equitable quality), education, empowerment of women and girls, management of water resources, accessibility of sustainable energy, sustainable economic growth, and decent work (Zavratnik, Kos, & Stojmenova Duh, 2018). This demonstrates the wide scope that can be tackled with what are considered smart approaches/initiatives.

The 'smart' concept is linked to innovation. Some scholars related it directly to the application and usability of information and communication technology (ICT) in the context of a village (Visvizi & Lytras, 2018) and others to innovative processes, services, and products introduced in a territory (OECD, 2018). Assuming this relation between 'smart' and innovation (including social and digital innovation), smart concepts are at the center of the new policy paradigm proposed by the Organisation for Economic Cooperation and Development (OECD). This key international organization explicitly acknowledges that innovation will be critical for the future competitiveness and sustainability of rural economies if it is embedded in rural policies which follow a *place-based* approach and rely on *multisector coordination* and *multi-level governance* to unleash growth potential that is grounded in rural-specific assets (OECD, 2018).

More specifically in the EU context, LEADER is an approach to policy implementation which has a long-track history spanning almost 30 years. Since 2007 it has been part of the EU Rural Development Policy framework. This integrated approach illustrates 'on the ground' that rural development policies, as indicated by OECD, need to widen their focus beyond economic development of the agricultural sector to include other aspects such as the quality of public services and new economic opportunities, for maintaining quality of life and attractiveness of rural areas both for people and for businesses (OECD, 2018).

'Smart villages,' however, is a recent concept, triggered by the EU Action for Smart Villages established jointly by three Directorate-Generals (DGs) of the European Commission (DG Agri, DG Regio, and DG Mobility) in April 2017, backed by strong commitment in the European Parliament. Since then, there is growing interest around the smart villages concept at all levels – policymakers and implementers at EU/national/regional/local levels. Part of this interest is due to its vision of balanced development in European regions and the need to provide growth and cohesion perspectives to rural areas and villages.

There is great interest at the political level around smart villages, illustrated by the number of policy initiatives/opinions developed on this subject. The European Economic and Social Committee (EESC) acknowledges the contribution that smart villages can make in addressing the challenges of rural Europe. It calls for wider support for these initiatives from various national and regional administrations, a call which outlines the wide scope of this concept (EESC, 2017). The Committee of the Regions (CoR) interconnects LEADER and smart villages to tackle modern problems in rural areas in Europe (CoR, 2017). It acknowledges that bottom-up approaches are key to implementing smart villages, and that LEADER – as a proven successful initiative – can play an important role. However, the opinion also emphasizes that implementation of smart villages cannot rely solely on LEADER because the problems of rural areas require a significant and wide-ranging number of responses. Instead it calls for additional rural actors and institutions to be involved in local development, such as local and regional authorities, including those from other EU funds. These opinions were followed up by the pilot project on smart eco-social villages, which intends to bring a common European definition to smart villages (European Commission, 2018), the Bled

Declaration and the European Network for Rural Development (ENRD) Thematic Group on smart villages (ENRD, 2017).

Initiatives to develop or implement smart village policy frameworks are also taking place in the member states. It should be noted that these policy initiatives are not always explicitly referred to or named as smart villages (ENRD 2018b). The ENRD explains that these initiatives are often driven by growing concern for the major challenges affecting rural areas such as *depopulation* and cuts to *public services* in rural areas (ENRD, 2018c). At the same time, there is increasing recognition that rural communities need to be enabled to make the most of new opportunities offered by digital transformation, the transition to a low carbon, circular economy, new forms of urban−rural linkages, emerging value chains, and others.

Two important elements emerge in the implementation of smart villages under the integrated rural development framework. First, it puts *people and communities at the center of development*, empowering rural communities to implement specific future-oriented innovative local initiatives that aim at the sustainable development of their area, hence, relying on a *bottom-up approach and cooperation*. Second, it is a *placed-based approach* which supports actions tailored to the local rural context in order to capture the specificities of the areas.

This is the case with the *National Strategy for Inner Areas* in Italy, or the *Plan for Rural Digitisation* in Finland. The Italian case represents one of the most comprehensive and integrated national strategies for tackling the problems of depopulation and reduced access to services in Italy. All four European structural and investment funds are combined with national finance to support strategies for both local development and service innovation in 72 pilot areas with a total budget of around 1 billion euros (ENRD, 2018d). In this plan, LEADER LAGs play a variety of roles ranging from the animation of the territory, facilitation of community processes, to project support.

The Plan for Rural Digitisation in Finland supports the development of digital services and promotes experiments and pilots in rural areas of the country. It aims to provide working conditions for rural citizens and businesses and create new vitality and growth prospects for the business community in rural areas. Interventions aim to develop extensive telecommunication networks at regional level, design accessible e-services, create the conditions for new rural mobility services, promote remote access, and support rural digital businesses. The Rural Development Programme finances initiatives under this plan, including those supported through LEADER (ENRD, 2018c) (Figure 1).

The LEADER approach and the smart villages concept are both embedded in the integrated development paradigm. Current integrated policy initiatives show that LEADER is one of the key players in some national policy frameworks set up to implement smart villages in the EU. This could be considered a natural reaction to the concept of smart villages as defined in the EU Action Plan. This definition for smart villages outlines specific features shared with the LEADER approach, such as *area-based development*

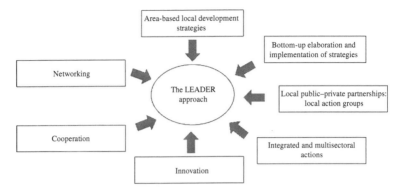

Figure 1. LEADER Principles. *Source*: Own elaboration.

process, cooperation and partnership, innovation, integrated focus, and *strategic implementation.*

Considering the principle that policies need to work in synergies, LEADER and smart villages therefore have a strong potential to be complementary, supporting each other to enable rural areas to address current challenges and to seize opportunities for sustainable development in areas such as digitization, social innovation, bioeconomy, energy, mobility, social care, environmental sustainability, and others. This is recognized in the policy debate, confirming that LEADER is an appropriate approach and tool to support smart village initiatives (Bogovič & Szanyi, 2018).

Smart Villages as an Opportunity to Strengthen LEADER

The previous analysis outlines that there are common features shared between smart villages and LEADER. Aspects such as (1) place-based approach, (2) integrated development (socioeconomic–environmental), (3) innovation, and (4) partnership cooperation establish common ground on which both LEADER and smart villages can complement each other. In addition to this, there are elements that can allow the two to strengthen each other. These are the result of their practical differences. Table 1 outlines some of these, and further explanation follows in the text of this section.

An analysis of the differences presented in Table 1 clearly shows that they emanate from the limitations posed on LEADER to be deployed to its full capacity. Most of these barriers arise from an inappropriate design of the regulatory framework under which LEADER operates in the member states. Considering the political enthusiasm for smart villages, and its close links with LEADER, an opportunity exists to boost the rejuvenation of LEADER in many member states, and to enable LEADER to do what it is meant to do.

Table 1. Distinctive Characteristics between Smart Villages and LEADER.

Features	Smart Villages	LEADER
Place/area-based approach	Focus on sub-local areas, villages, or groups of village communities	Focus more on a larger territorial level than single villages/communities in their LDS
Integrated approach	Focus on a vertical integrated approach with renewed attention on rural services (education, health, mobility, environment, social capital, etc.). It aims to integrate local contexts/environments globally	Focus on a horizontal integrated approach, ensuring synergies among the various sectors of the territory
Partnership and cooperation	Focus on partnerships along a specific value chain or field to implement the smart village plans, including actors outside the local areas (e.g., academia and research)	Focus on transversal and territorial partnerships
Innovation	Renewed attention to local social and digital innovation in the wider rural sector (in addition to agriculture)	Focused on the areas limited by the regulatory framework

Source: (ENRD, 2018e) [16]

In the following text, we elaborate further on the common genuine characteristics of both concepts and their precise differences.

The Place/Area-based Approaches of Smart Villages and LEADER

With both smart villages and LEADER, initiatives and policies are tailored to the local rural context in order to capture the specificities of the areas. In the LEADER philosophy, the area-based approach takes a small, homogenous, socially cohesive territory, often characterized by common traditions, a local identity, a sense of belonging, or common needs and expectations, as the target area for policy implementation. Having such an area as a reference facilitates the recognition of local strengths and weaknesses, threats and opportunities, endogenous potential, and the identification of major bottlenecks for sustainable development.

Area-based essentially means local. This approach is likely to work better than other approaches because it allows actions to be tailored more precisely to suit real needs and local competitive advantage. The chosen area must have enough coherence and critical mass in terms of human, financial, and economic resources to support a viable local development strategy. It does not have to correspond to predefined administrative boundaries. The definition of a 'local area' is neither universal nor static. On the contrary, it evolves and changes with broader economic and social change, the role of farming, land management and environmental concerns, and general perceptions about rural areas.

However, smart villages aim to enable rural communities to tackle challenges and opportunities emerged locally by exploiting their local assets. It puts the focus on the villages/community level, in other words at a lower territorial level than LEADER does. Actions are based on a shared long-term vision with respect to one specific challenge or opportunity present locally. While smart villages call for a specific plan of action in one specific area (e.g., digitization, bioeconomy, basic services, mobility, tourism, etc.), LEADER designs an integrated (multisectoral) local development strategy (LDS) for the territory which can act as a higher-level territorial planning tool.

Integrated Development

In relation to integrated development, a differentiation is needed between a 'horizontal' and a 'vertical' integrated approach. LEADER focuses on the horizontal integrated approach, drawing local development strategies on the basis of a territorial analysis of the social economic and environmental problems. Through such a strategy, individual projects are financed with more of a focus on private projects and in many cases prioritizing job creation. This is the result of the demanding EAFRD regulatory framework and indicators that puts LEADER under policy objectives and targets related to job creation.

However, LEADER is not a sectoral development program; local development strategies have a multisectoral rationale, integrating several sectors of activity. The actions and projects contained in local strategies are linked and coordinated as a coherent whole. Integration may concern actions conducted in a single sector, all program actions or specific groups of actions or, most importantly, links between the different economic, social, cultural, environmental players and sectors involved. Stronger attention is paid to links and connections within the local territory while less attention to enhancing those with the global challenges and actors.

Smart villages apply a vertical integrated approach with a strong thematic focus. Initiatives supported have a specific medium- or long-term plan that supports a series of actions in a specific area/field (e.g., energy, agriculture, mobility or care, etc.) to achieve the planned objectives and which benefit the community. Smart villages are about guiding the energy and vision of local people and communities toward local actions and making links with global challenges and opportunities.

The project examples and initiatives that have been collected and discussed in the ENRD Thematic Group on smart villages clearly indicate that smart villages usually begin with local people coming together around a common problem or a common vision in order to implement some form of 'plan of action' to achieve a specific goal. Initiatives design packages of action that follow a development path to achieve the objectives (e.g., starting with animation, testing proposals, feasibility study, investments, marketing; see Figure 2).

In this way, rather than being another layer of comprehensive territorial document, smart village plans are simple plans for guiding and supporting collective action around a specific challenge or opportunity. Hence, smart villages emphasize community/collective actions more than individual ones (ENRD, 2018e).

Partnership and Community Empowerment

The LEADER local development strategies are elaborated and implemented by local actions groups (LAGs), a formal partnership that integrates the key stakeholder groups of the territory. The LAGs are multisectoral and the decisions are taken 'bottom-up.' LEADER is an approach that empowers local communities to be at the center of their development by supporting them in specific projects.

Smart village plans also try to support local communities' actions through a specific plan of action. These plans are not about preparing another sort of participatory planning process which is then implemented through a formal program/strategy (as in the case of LEADER). Instead of such a comprehensive approach, they tend to start small and focus on resolving certain key opportunities or problems that motivate local communities to change. To do so, they require cooperation between different agents within the territory and beyond (e.g., research and academia) to implement the action plan in a coordinated manner, so that it benefits the community as a whole. This more flexible approach allows for all relevant stakeholders to engage in development, including emerging stakeholder groups who are in many cases not often represented in LEADER partnerships, such as academia and ICT. Hence, these plans link local level actors with other actors who can create local benefits.

Figure 2. Potential Actions in Smart Village Plans. *Source*: ENRD (2018m).

Innovation

Innovation already stands as a top priority in the European Rural Development Policy framework. This gives considerable support to innovation in the agricultural sector by implementing European Innovation Partnership (EIP) operational groups (EIP-AGRI, 2018).

LEADER, since its beginning, was conceived as a key mechanism for boosting innovation in rural areas beyond the agricultural sector. Since having been mainstreamed into the rural development framework in 2007, innovation possibilities have become less frequent (ENRD, 2010, 2018f). Even with innovative objectives in the local development strategies, projects often have to abide by the eligibility framework of standard predefined measures in the relevant EAFRD program, and this has constrained innovation elements. The regulatory framework does not give sufficient room to implement risky innovative projects, leaving the innovation niche without a policy tool. The ENRD Survey on LEADER outlines some worrying conclusions on this constraint for innovation in LEADER (ENRD, 2018a):

- Implementing cooperation projects and innovative approaches are somewhat challenging for LAGs, particularly for LAGs with small budgets.
- LAG staff spends the most time on 'Supporting project development,' but given the choice, they would like to spend more time on project development, animation, and innovation. It appears that the things that 'have to be done,' for example, administration and management, take too much time away from longer-term development activity and objectives, such as building community capacity and fostering innovation.
- Finding innovative solutions to local problems is challenging for LAGs while they see this as very important for rural areas.
- Overall, there are differences in the implementation of LEADER among countries, and in those where difficulties are greater, the implementation of innovation is more challenging.

The smart villages concept outlines the need to boost innovation in all relevant areas. It puts innovation at the center, supporting new solutions that help rural communities overcome specific challenges or seize opportunities in all relevant rural sectors such as agriculture, public services, energy, healthcare, education, and mobility among others. The concept also goes beyond standard measures of local development by enabling local people to test out and implement new solutions to the problems they face.

Therefore, while innovation in the agricultural sector is supported by EIP groups, it remains an orphan in the case of specific policy tools for wider rural issues (see Figure 3). The smart villages concept brings this gap in the rural development agenda to light. It is a gap which could be filled by LEADER (if LAGs are given the right regulatory framework). As a result, the smart villages concept provides an opportunity to refocus innovation toward other rural sectors in addition to agricultural innovation. The EIP-AGRI model is already well-known by public authorities and standardized in public administrations

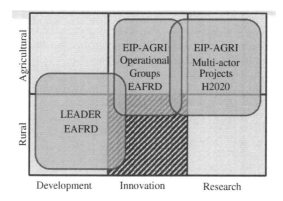

Figure 3. Mapping Rural Innovation and Policy Tools. *Source*: Pertoldi, Muriel, and Lopez (2016).

systems. Extending this model of innovation to wider rural development issues could be an effective way to fill this innovation gap.

The Potential Role of LEADER in Smart Villages

In the 2014–2020 period, LEADER has a budget of €9.8 billion public expenditure (EAFRD and national funding) and represents 6% of the planned total RDP expenditure. There are 2,562 LAGs foreseen in the EU for the 2014–2020 period. It is important to note the geographical scope of LEADER in four countries – France, Germany, Poland, and Spain – where there are more than 200 LAGs per member state. This enables LEADER to be best placed to *facilitate a quick and efficient rollout* of the smart villages concept across the EU. LAGs can be an invaluable initiator, relay or multiplier in supporting smart villages.

The LEADER measure in the RDPs has the potential to be the most *versatile funding source for smart villages*: LDS developed by the LAGs can include specific support to smart village plans to cover several fields of intervention, such as energy, mobility, care, territorial inequalities, or climate action. However, this is not the norm across all EU LAGs for the following reasons:

- lack of the resources in LAG teams to engage expertise in these new areas. LAGs composed of broad partnerships can sometimes overcome this constraint, although often partnerships do not include stakeholders relevant in new emerging fields such as innovation, digital, energy, climate, and mobility among others; and
- limitations placed by specific national regulations for LEADER in some countries.

When managed correctly, LEADER funding can be used to *provide integrated support itineraries* that can take smart community initiatives from their original idea through the experimentation stage all the way to launch. This can complement the traditional work of the LAGs in supporting individual projects to address short-term needs – which are highly important in the rural context – with work on smart village plans which have long-term vision.

Individual LAG budgets vary substantially across EU member states. They range from less than €1 million over the whole period, to over €9 million (in Greece or Ireland) and up to €15 million in Saxony (Germany) where some 40% of the RDP is implemented through LEADER. LAGs with smaller budgets and restrictive financial options are best placed to mobilize and animate communities to develop smart village plans, while investment actions can then be funded by sources other than the LEADER budget.

However, at the other end of the scale, LAGs with significant budgets have the funds to invest directly in the smart village plans, including *small-scale infrastructure and services*. In addition, as part of the support for smart village plans, LAGs could *pilot innovative projects* and carry out *feasibility studies* with little cost incurred. LAGs have long-standing experience implementing 'own projects' that would allow for this experimentation.

Funds available to LEADER alone are normally insufficient to address all the investment needs faced by rural areas. It is only when they are combined with other RDP measures and other funds (e.g., cohesion funds, financial instruments, H2020, etc.) or when several LAGs join forces to cooperate that they can become a more powerful force.

LEADER can play various roles in implementing the smart village concept. These roles are analyzed below by looking specifically at projects identified by the ENRD as good examples of smart village initiatives (ENRD, 2017). Given the wide scope of this concept, the selected projects cover a variety of sectors such as bioenergy, mobility, food and agriculture, entrepreneurship and training, and broadband. Lastly, all projects also share the fact that LEADER has been in the driving seat for their implementation. This allows a systematic analysis of the genuine role of LEADER in smart villages.

LEADER as Animator and Facilitator of Community Processes

This has been one of the most important features of LEADER and one which brings an ultimate value to the rural context in terms of social cohesion. Animation and facilitation of communities to design and implement smart village plans is an essential step. Cooperation processes often emerge spontaneously in rural areas characterized by strong leadership, social capital, and a tradition of working together. But since this is not the case everywhere, the animation and facilitation of those communities seem to be the key to ensure they take advantage of these opportunities. This role can be carried out by LEADER LAGs and is not new to them.

This has been the case in many parts of Europe and can be exemplified by the Bioenergy Villages in Germany. The project seeks to promote local renewable energy production in the villages in the defined area. With five Bioenergy Villages now operational, the project connects local farmers to village cooperatives which manage energy production and distribution (ENRD, 2018g). The project attracted high levels of local citizen participation and provides important social, economic, and environmental benefits. LEADER supported with initial funding was mainly used to conduct the feasibility study. Complementary funding was provided by other sources from the local community itself and the federal government.

In this case, LEADER helped capitalizing the bottom-up initiatives and brought together various stakeholder groups with a common interest and vision in a particular field (energy). It animated and facilitated a quite complex, resource-intensive process. In addition, LEADER supported specific steps within the plan of the Bioenergy Villages which were particularly aimed at reducing the risks of this innovative project by conducting a feasibility analysis. This 'non-material aspects' or 'soft' contribution helped to ensure the success of the project. The case also suggests that it seems not be in the remit of the LEADER LAG to support larger investment projects with LEADER funding, considering the financial limitation LAGs have.

LEADER as a Laboratory of Innovation for a Transitional Change

The implementation of new solutions often implies a higher degree of risk and uncertainty to ensure the success of innovative initiatives. This is an intrinsic feature of the innovative projects which public policies must support for innovation to happen in rural areas. However small steps exist which can bring more certainty to future outcomes.

An example of this is found in the project 'building a food ecosystem' which was implemented by the 'Pays de Condruses' LAG, Belgium (ENRD, 2018h). With the support of LEADER, this local area implemented a series of innovative initiatives aimed at creating and interconnecting an environment through food. It supported a cooperative of food producers, capacity building actions, incubators for market gardeners, and food hubs of the area, actions which collectively helped to build an innovative territorial ecosystem based on food. Through LEADER cooperation, the project scaled-up from a local initiative and extended to form a network of food incubators in Wallonia (Belgium).

LEADER acted as a laboratory for innovation. Mobilizing, animating, and engaging the communities and villages is an arduous and hidden task and one which was indispensable for this project. LEADER was necessary to trigger innovative components of the project. It showed that it can have a pivotal role in bringing different stakeholders together, animating and facilitating the innovative processes and initiatives.

LEADER Supporting the Scaling-up of Initiatives through Cooperation

While smart village initiatives often start small, some aspire to be implemented on a wider scale. Cooperation has been seen to be key to allow small-scale pilot initiatives to achieve a greater dimension – once they have demonstrated the benefits they bring at a smaller scale. Cooperation has been at the core of LAG activities since the beginning of this method in the 1990s. LEADER has already experience and – most importantly – the network and resources to enable cooperation locally, regionally, nationally and internationally.

Interregional cooperation was the key aspect that enables the scale-up of a local initiative implemented in Catalonia (Spain) on coworking spaces in rural areas. The project 'COWOCAT,' initiated by four LAGs from the region, aimed to create a single coworking platform to provide spaces to stimulate economic activity and help maintain the population in rural areas (ENRD, 2018i). Following a pilot project to set up coworking spaces in rural areas, 10 Catalonian LAGs launched a continuation interterritorial cooperation project to create and expand the network and promote this new way of working more widely in rural areas. Over a two-year period (2014–2016), some 14 coworking spaces were created, hosting 60 to 65 coworkers each, connecting more than 130 professionals with the coworking spaces. The local entrepreneurs use the place not only to work but also to stimulate cooperation projects among the local community, create synergies, and try to attract new business. More recently, this project has achieved a more European dimension through LEADER cooperation, and is now being implemented by LAGs in France, Lithuania, United Kingdom, Ireland, and Germany.

Another example is the project Rezo Pouce (France) that, with the support of LEADER, tackled mobility issues in rural areas by providing a safe and free way for people to car share on short journeys between key locations (ENRD, 2018j). The Rezo Pouce enables carpooling at designated hitch-hiking spots. Registered users can find a ride to and from their chosen destinations. The subscription fee for the municipality is based on the number of registered users of the service. For municipalities with a population of 10,000 to 25,000, the fee amounts to €7,500 and €3,000 for two years. In return, the social enterprise behind Rezo Pouce supports the rollout and provides technical and coordination assistance. It also trains dedicated municipality staff charged with overseeing the scheme. LEADER has supported the scale-up of this initiative into other municipalities.

Those examples on top of others introduced earlier illustrate how LEADER LAGs have reacted to new challenges in their areas in order to find smart solutions in a way that is typical for LEADER: pilot projects which are rolled out and scaled-up, often with the help of cooperation.

LEADER as an Enabler for the Transition to New Innovative Fields Such As the Digital Economy

The digital economy is a new emerging opportunity for rural areas. Interventions in this new innovative area tend to be applied to more urbanized territories and fall outside the traditional scope of intervention of LAGs in rural areas. However, smart communities are taking steps to change this, and LEADER has an incredible role to play in addressing the three core dimensions for digitization: (1) broadband infrastructure, (2) skills and capacities, and (3) digital innovation.

Broadband infrastructure is the basic element which ensures that rural communities can take advantage of the emerging opportunities from the digital economy. Yet in many cases, these areas remain white spots for broadband because the private sector does not consider them attractive enough. Smart communities do not wait for broadband to arrive. By mobilizing and aggregating local demand for broadband, interest can be sparked in the private sector. This was the case in 'The Balquhidder Broadband' project in Scotland. In order to create the necessary infrastructure, the community approached the community interest company "Balquhidder Community Broadband (BCB)" (ENRD; 2018k). By creating a community interest company which owns the network, genuine collaboration has emerged between local businesses, local government, local people, and the commercial partner Bogons so as to lay the foundations for broadband connectivity in Balquhidder. It now delivers hyperfast, future-proof, and community-owned broadband to all 197 homes and premises in the Balquhidder area which are then better able to access commercial and public services online. Stirling Council and Forth Valley LEADER LAG jointly fund the project with a Project Officer. Once more, LEADER was the driving force for bigger investment and helped to match public and private partners to achieve smart cooperation in an area that falls outside LEADER's traditional scope of action.

The availability of broadband infrastructure does not guarantee that it is used to its maximum potential in rural areas. Digital skills and capacities are key to foster digital innovation. That was exactly the case in Pays du Grand Bergerac in France. Digital technologies were available in the area but remained unused. LEADER supported a digital audit in the LAG area. These audits are a free, confidential, and tailored report that provides a comprehensive analysis of digital status and makes early recommendations for digital business development. Audits were sent to 120 rural SMEs, sparking the interest of nearly 50 business managers to attend training courses on the key aspects digital tools can play in their business development. These activities were part of the 'WAB' initiative (ENRD; 2018l), a hub for innovation and digital training operating in rural Bergerac. The training which was available for local businesses offered by local authorities in the area of Dordogne did not focus enough on digital aspects. In 2016, a new social enterprise was created to help change this. As a social enterprise, WAB aimed to boost digital use among small rural businesses and aspires to install a digital city in Bergerac. This is known as a 'web school' that teaches local businesses to design effective strategies that incorporate digital aspects in organizing work, communication, and marketing. With more than 200 businesses supported in two years of implementation, the project enabled the creation of spin-offs in the form of an enterprise network and the development of a coworking space. In addition, it provides ICT training for unemployed people to support local businesses.

Conclusions and Recommendations

At first glance it seems that the introduction of the smart villages concept into the proposals of the European Commission for the Rural Development Policy post 2020 does not explicitly reflect the potential of LEADER and its contribution to the use of smarter approaches in rural areas. This reflection appears critical because member states will explore the best way to implement smart villages in their own national contexts. This chapter presented some thoughts to be considered when designing the policies for smart villages and LEADER.

One key conclusion that emerges from the above analysis is that LEADER and smart villages are conceptually closely related. As a consequence, LEADER seems a key policy approach and tool to implement the smart village concept in practice in the near future. Individual actions to support smart village plans can be integrated under Local Development Strategies, forming a specific type of intervention to be supported. However, this chapter also outlines that the way LEADER is designed in the policy framework has the potential to pose some limitations in how it can be applied to its full potential, and ergo to the implementation of the smart villages concept. Eliminating these regulatory barriers in LEADER can unlock the potential of smart villages in rural areas. To address this, some of the following recommendations are made:

- The smart villages concept focuses on the transition toward a new rural model based on communities' long-term vision. Regulations should be flexible enough to allow LAGs to work on this long-term transition process on top of addressing the short-term needs of the area. Additional resources must be dedicated to LEADER if it is aimed to implement the smart villages concept beyond the indispensable work it currently performs at local level. While smart villages cannot substitute the support LAGs currently provide on the ground to individual small projects, it can build on the existing work. Displacement of current funding must be avoided.
- LAGs should have at their disposal more resources dedicated to mobilizing, engaging, and facilitating the smart village processes. The smart villages concept puts its focus on community and cooperation processes, and these are quite resource-intensive in both economic and time terms. LAGs need to be able to accompany villages/communities through a long-term process, particularly important for some communities to ensure that they do not become excluded from these opportunities. A suitable framework for implementing multifunded CLLD offers a good potential for smart villages. It is also key to put in place flexible regulatory frameworks in other policy supporting tools from the European Structural and Investment Funds (ESIF) to draw additional resources to implement smart village initiatives. Examples of these additional tools are Integrated Territorial Instruments (ITIs), cooperation interventions, CAP digital strategies, or investment interventions within the CAP.

- Smart villages can potentially be applied in any sector that generates development interest in local communities. LAGs have not traditionally been involved in some of these newer sectors, such as digitization, mobility, health, or education. But in others, LAGs have accumulated a lot of experience throughout the years such as agriculture, value chains, tourism, and environment. This broader scope of interventions poses a challenge for LAGs to build new expertise in a wider set of fields, something which requires them to broaden their knowledge base. LAGs resources should be sufficient to be able to acquire, recruit, train, and upskill their teams in order to achieve the skills/knowledge needed. In addition, LAGs should rethink their partnerships, building links with partners in new areas such as digitization, energy, and climate, to support the process. These could be from sources outside their working territory. The future LDS should help identify those areas relevant to the implementation of smart villages, including in the new emerging fields.
- To ensure the rollout of the smart villages concept through LEADER, regulation must enable LAGs to provide simplified access to global grants to support small-scale smart village plans (covering 'soft' and 'hard' actions such as animation, feasibility studies, piloting, small-scale investment, and promotion) or to provide support to specific activities of a plan. Moreover, funding requests coming from the smart village processes could be prioritized in other RDP supporting tools (e.g., cooperation, investment, training, advisory services, etc.) or in other ESIF (ESF, ERDF, etc.).

References

Bogovič, F., & Szanyi, T. (2018). *Smart villages are modernisation tools to be taken seriously*. Euractiv, September 24. Retrieved from https://www.euractiv.com/section/agriculture-food/opinion/tue-smart-villages-are-modernisation-tools-to-be-taken-seriously/?_ga=2.156877243.1281855723.1537885384-783752980.1537885384. Accessed on September 25, 2018.

CoR – European Committee of the Regions. (2017). *Opinion on revitalisation of rural areas through smart villages*. Retrieved from https://cor.europa.eu/en/our-work/Pages/OpinionTimeline.aspx?opId=CDR-3465-2017.

EESC. (2017). *Villages and small towns as catalysts for rural development*. European Economic and Social Committee. Retrieved from https://www.eesc.europa.eu/en/our-work/opinions-information-reports/opinions/villages-and-small-towns-catalysts-rural-development. Accessed on September 25, 2018.

EIP-AGRI. (2018). European innovation partnership for agriculture and innovation. Retrieved from https://ec.europa.eu/eip/agriculture/en/about. Accessed on September 25, 2018.

ENRD - European Network for Rural Development. (2018a). *LEADER Survey 2017*. Retrieved from https://enrd.ec.europa.eu/sites/enrd/files/leader-resources_lag_survey_results.pdf. Accessed on December 9, 2018.

ENRD. (2010). Leader subcommittee Focus Group on preserving the innovative/experimental character of LEADER: Extended report (November 2010). European

Network for Rural Development. Retrieved from https://enrd.ec.europa.eu/sites/enrd/files/fms/pdf/7CC8B99E-925C-7C33-CE76-1814EDEAD8A3.pdf. Accessed on November 24, 2018.

ENRD. (2017). *Smart villages portal*. European Network for Rural Development. Retrieved from https://enrd.ec.europa.eu/enrd-thematic-work/smart-and-competitive-rural-areas/smart-villages_en. Accessed on September 25, 2018.

ENRD. (2018b). *Rural review on smart villages: Revitalising rural services*. European Network for Rural Development. Retrieved from https://enrd.ec.europa.eu/sites/enrd/files/enrd_publications/publi-enrd-rr-26-2018-en.pdf. Accessed on September 2018.

ENRD. (2018c). *Scoping paper on smart villages*. European Network for Rural Development. Retrieved from https://enrd.ec.europa.eu/sites/enrd/files/tg_smart-villages_case-study_it.pdf. Accessed on September 25, 2018.

ENRD. (2018d). *Strategy for inner areas in Italy, briefing document*. European Network for Rural Development. Retrieved from https://enrd.ec.europa.eu/sites/enrd/files/tg_smart-villages_case-study_it.pdf. Accessed on September 25, 2018.

ENRD. (2018e). *Smart villages and LEADER.* 6th ENRD Thematic Group meeting, presentation by Eamon O'Hara, European Network for Rural Development. Retrieved from https://enrd.ec.europa.eu/sites/enrd/files/tg6_smart-villages_leader_ohara.pdf. Accessed on December 14, 2018.

ENRD. (2018f). *Innovation in the LEADER delivery chain – A summary based on the LEADER innovation PWG discussions and meetings*. European Network for Rural Development. Retrieved from https://enrd.ec.europa.eu/sites/enrd/files/leader-innovation_delivery-chain_pwg-discussions.pdf. Accessed on November 22, 2018.

ENRD. (2018g). *Bio-energy village*. ENRD Smart Villages Seminar 2018, European Network for Rural Development. Retrieved from https://enrd.ec.europa.eu/sites/enrd/files/s7_smart-villages_bioenergy-village_de.pdf. Accessed on September 25, 2018.

ENRD. (2018h). *Presentation building a food ecosystem*. ENRD seminar on LEADER: Acting Locally in a Changing World, European Network for Rural Development. Retrieved from https://enrd.ec.europa.eu/sites/enrd/files/s9_leader-seminar_lags_pays-de-condruses_be.pdf. Accessed on December 9, 2018.

ENRD. (2018i). *'COWOCAT-RURAL' promoting coworking in rural Catalonia*. ENRD projects database, European Network for Rural Development. Retrieved from https://enrd.ec.europa.eu/projects-practice/cowocatrural-promoting-coworking-rural-catalonia_en. Accessed on September 25, 2018.

ENRD. (2018j). *Rezo Pouce, mobility*. ENRD Smart Villages Seminar 2018, European Network for Rural Development. Retrieved from https://enrd.ec.europa.eu/sites/enrd/files/s7_smart-villages_rezo-pouce_fr.pdf. Accessed on September 25, 2018.

ENRD. (2018k). *Balquhidder broadband*. ENRD Smart Villages Seminar 2018, European Network for Rural Development. Retrieved from https://enrd.ec.europa.eu/sites/enrd/files/s7_smart-villages_balquhidder_scot-uk.pdf. Accessed on September 25, 2018.

ENRD. (2018l). *La Wab*. ENRD Smart Villages Seminar 2018, European Network for Rural Development. Retrieved from https://enrd.ec.europa.eu/sites/enrd/files/s7_smart-villages_la-wab_fr.pdf. Accessed on September 25, 2018.

ENRD. (2018m). *Smart villages: Results from the ENRD Thematic Group. Presentation by Paul Soto at the fi-compass workshop EAFRD Financial Instruments Working with

Rural Infrastructures, European Network for Rural Development. Retrieved from https://www.fi-compass.eu/sites/default/files/publications/20180619_brussels_Paul%20Soto.pdf. Accessed on January 14, 2019.

European Commission. (2006). *The LEADER approach: A basic guide*. Luxembourg: Office for Official Publications of the European Communities.

European Commission. (2017). *EU action for smart villages*. Retrieved from https://ec.europa.eu/agriculture/sites/agriculture/files/rural-development-2014-2020/looking-ahead/rur-dev-small-villages_en.pdf. Accessed on September 25, 2018.

European Commission. (2018). *Pilot project smart eco-social villages agri-2017-eval-08*. Retrieved from http://www.pilotproject-smartvillages.eu/. Accessed on September 25, 2018.

European Commission, European Parliament. (2018). *Bled Declaration for a smarter future of the rural areas in EU*. Retrieved from http://pametne-vasi.info/wp-content/uploads/2018/04/Bled-declaration-for-a-Smarter-Future-of-the-Rural-Areas-in-EU.pdf. Accessed on September 25, 2018.

OECD. (2018). *Edinburgh policy statement on enhancing rural innovation*. OECD. Retrieved from http://www.oecd.org/cfe/regional-policy/Edinburgh-Policy-Statement-On-Enhancing-Rural-Innovation.pdf. Accessed on September 25, 2018.

Pertoldi, M., Muriel, M., & Lopez, J. (2016). *Smart LEADER: Challenging smart specialisation in the scope of rural development. Presentation at the conference on Smart Specialisation and Territorial Development*. Retrieved from https://3ftfah3bhjub3knerv1hneul-wpengine.netdna-ssl.com/wp-content/uploads/2018/07/Pertoldi_SmarLEADER_final.pdf. Accessed on January 14, 2019.

Visvizi, A., & Lytras, M. D. (2018). It's not a fad: Smart cities and smart villages research in European and global contexts. *Sustainability*, *2018*(10), 2727. doi:10.3390/su10082727

Zavratnik, V., Kos, A., & Stojmenova Duh, E. (2018). Smart villages: Comprehensive review of initiatives and practices. *Sustainability*, *2018*(10), 2559. doi:10.3390/su10072559

Chapter 6

Precision Agriculture and the Smart Village Concept

Daniel Azevedo

Introduction — What Is the Role of Agriculture in Rural Areas?

Stakeholders all over the European Union (EU), including academics and the European institutions, are interested in sustaining a better life for rural area inhabitants. Current and future challenges faced by rural areas across the EU have been identified in several documents, such as the Cork 2.0 Declaration or communication on the future of food and farming, and others.

The pleas entailed in this kind of documents and declarations build on the legacy of the Treaty of Rome and the Common Agriculture Policy (CAP). The CAP was established to provide affordable food for EU citizens and, at the same time, a fair standard of living for farmers. Owing to the post-war circumstances in which it developed, key in CAP is the principle of food security, which remains extremely relevant today. As such it underpins the logic behind the EU-level debate on rural areas and, hence, smart villages.

But most important, they all agreed on the key role of agriculture for the future of rural areas. In 2017, agriculture and forestry cover more than 75% of the land. Agriculture and forestry have provided huge contributions to the culture, landscape, management of the natural resources, and the economy of rural areas — it has been a keystone for sustainability of rural areas. It has transformed European landscapes since people were able to learn how to use the local resources in order to transform the landscape, whether by setting fields on fire to clear the ground of pastures, building artificial dams to store water, or by favoring different varieties of plants in order to ensure supply of food.

The result of human activity is a variety of landscapes that are directly dependent on activities related to farming (crops, forestry, hunting, etc). There are also a wide variety of activities directly and indirectly linked to agriculture and food production (e.g., hunting and aquaculture). We also see a wide variety of systems of production in order to match consumer demand (organic, free-range, extensive, intensive, etc.). Never in the history of mankind, European and world consumers could access such safe, affordable, and nutritious food produced according to the highest safety, environmental, labor, animal health, and welfare standards in the world.

Today, the EU agri-food sector is not only the backbone of the EU's rural areas but it is also a driver of the economy, delivering 44 million jobs in the EU, food security for 500 million consumers and representing 3.7% of EU GDP (European Commission, 2017). The EU agricultural and food production sectors are also densely integrated into the global market. The EU is the world's number one exporter of agricultural and food products[1] contributing with about €21 billion of the EU's positive trade balance in 2017, that is, around one-third of EU's total net trade balance. The EU is also the number one importer of agricultural products, in particular rare commodities, such as coffee and cacao.

From a different angle:

> Farmer's income has decreased by 20% in the last 4 years and is currently only at the level of 40% of the other sectors' income. European farmers' economic viability and competitiveness is crucial so that they can deliver on the environmental and social sustainability dimensions. (CEMA, CEJA, EFFAB, FEFAC, CEETTAR, ESA, Fertilizers Europe, & ECPA, 2018)

Between 2005 and 2013, the number of farms in Europe had been shrinking at an average annual rate of 3.7%, thus leading to consolidation, increases in competitiveness, and professionalization of the sector never seen before (European Commission, 2015). In some agri-food sectors, the integration of different stages of production allowed to reach impressive levels of efficiency.

The rate of technology uptake by the farming community, but also by the rest of the agri-food chain, has been increasing. Supply chains are more interconnected and integrated than ever before, and technology is expected to take a more active role. This has been clearly demonstrated by the 'Brexit' debate, where the parties recognize in the Political Declaration on the framework for the post-Brexit relationship agreed by the United Kingdom and EU that due to the:

> The Parties recognise that they have a particularly important trading and investment relationship, reflecting more than 45 years of economic integration during the United Kingdom's membership of the Union, the sizes of the two economies and their geographic proximity, which have led to complex and integrated supply chains. (European Union, 2018)

Agriculture and food production are also facing other tremendous challenges. These include limited resources, market choices, climate change, urbanization, and sometimes rural exodus also. Climate change is already showing its impact

---

[1] EU exports of agri-food products exceeded €138 billion in 2017.

on agriculture and food production by bringing additional unpredictability to sectors that depend on the 'mood' of natural elements.

European Commission's Communication and detailed analysis presenting its long-term climate vision for 2050 acknowledges that agriculture together with forestry "are two of the few sectors able to deliver solutions for removing Green House Gases (GHGs) from the atmosphere while reducing their own emissions on the long term" (European Commission, 2018a, 2018b). Some of the 20% reduction of emission carried by EU farmers in the last years, while increasing production, was built due to the uptake and implementation of sophisticated information and communication technology (ICT) in farms. Advances in ICT, including the Internet of Things (IoT), automation combined with artificial intelligence (AI), or new breeding technologies, will be crucial for agriculture and forestry. These technologies will provide tools necessary to address challenges the sectors face, as well as to deliver a positive contribution to fight climate change.

Since the dawn of agriculture, farmers have used the most advanced technology and the most advanced knowledge available to them to take the best decision possible, at that particular moment. Over the centuries, agriculture went through three major technology transformations. The first stage, it was an agriculture based on intensive labor (main employer of the population) with high number of small farms but low productivity. It employed a huge number of the population. In the beginning of the twentieth century, Agriculture 2.0 kicked in with the advent of mechanization. Agriculture 3.0, also known as 'Green Revolution' started around the 1960s with the introduction of fertilizers and synthetic pesticides that allowed a huge increase in productivity.

Today, agriculture, as most of the economic sectors, has embraced a new technological transformation, stepping forward to Agriculture 4.0! And believe it or not, agriculture is a frontrunner in this technological transformation. That being said, the question is how agriculture and food production can contribute to and influence the policy strategies related to smart villages? This chapter dwells on this issue. To this end, in the first part, the concepts of precision agriculture and farming are introduced, including the new trends of technology applied in the agri-food sector. In the next step, the importance of agriculture and food production in the smart villages concept is examined. Then, the key bottlenecks are discussed. Conclusions and recommendations follow.

What Is Precision Agriculture/Farming?

Precision agriculture (PA) or precision farming (PF) is a farming management concept based on observing, measuring, and responding to inter- and intra-field variability in farming. Considering the content of the entire volume, it is meaningful to highlight that there is a big debate about the use of the terms 'precision' agriculture and 'precision' farming as contrasted to 'smart' farming. Even if the difference may not be big, in this chapter the term precision agriculture will be employed, because only by getting more precise, farming can become 'smarter.'

From a different angle, the term 'smart' might be wrongly interpreted as a suggestion that, currently, agriculture is not smart; but that is obviously not the case.

What are the factors that render agriculture becoming more precise? With the uptake of AI systems, should we start naming it 'intelligent agriculture'? PA provides the farmer with decision support systems that allow primarily better management of all aspects of the farm. The key word is precision. The precision is provided by the ability to collect and process huge amounts of data, and extract information from them. Indeed, the onset of the big data paradigm (Lytras, Raghavan, & Damiani, 2017) and data-driven decision-making may have immense implications for farming also.

PA means that:

> plants (or animals) get precisely the treatment they need, determined with great accuracy thanks to the latest technology. A range of forms of technology are used to this end, including GPS, sensor technology, ICT and robotics. Technology can assist in strategic decision-making at farm level as well as with operational actions at plant level. This allows production to be optimised and means we can work on more sustainable crops. The big difference with classical agriculture is that rather than determining the necessary action for each individual field, precision agriculture allows actions to be determined per square meter or even per plant. (WUR, 2018)

This means that the farmer is able to match his/her farming practices to the exact needs of the plants and animals of his/her farm. The efficiency on the use of resources translates into additional competitiveness and reduction of environmental risks and footprint.

PA is not a new concept. The use of satellite and satellite imagery in the 1970s and global positioning system (GPS) in the 1990s in agriculture, that allowed to locate the precise position in the field, has enabled farming to make substantial steps toward a more controlled and precise activity.

In recent years agriculture has seen many breakthroughs in several domains, such as biotechnology, nanotechnology, IoT and sensors, mapping, unmanned aerial vehicles (e.g., drones), remote sensing and satellite navigation, automated vehicles, and software. But it is the uptake of digital technologies that enable PF to aim for very ambitious targets and fulfill most of its promises. Today, agriculture and food production are taking up new technologies at a rate never seen in history, moving from intuitive decision-making to a technological and data-driven decision-making.

At the driver seat is big data. Big data is basically the precision word in PA. It refers to the process of collecting vast volumes of data from a variety of different data sources, analyze it, and obtain information that will primarily support the decision-making of the farmer. Farmers can upload the data collected from their farm, maps, and sensors onto data platforms (e.g., Farmbench), combine these data with other sources of data (e.g., input suppliers) through different algorithms/applications, benchmark with other farmers, produce agronomic, environmental or financial reports, and 'what if' scenarios of their business plans. This concept helps the farmer to move from intuitive farming to a data-driven agriculture, enabling to use the right amount of input in the right place at the right time.

In a report:

> Rabobank estimated that adopting these new practices can easily add USD 10 billion per year to the value of field crop farming on a global scale (Smit, 2015). In field crop farming, we estimate that adopting data-intensive farming practices could generate an additional value of about USD 10 billion per year, for farmers worldwide. This number is based on an estimated 5 percent yield increase on 80 percent of the area for the top seven crops (corn, soybeans, wheat, cotton, rapeseed, barley and sunflower). The real value will be higher, considering similar benefits to smaller high-value crops, such as sugarcane, potatoes, sugar beats, as well as fruits and vegetables. The livestock industry will also see similar benefits. (Smit, 2015)

These data can also be shared with the different partners of the agri-food chain, ICT community, or other stakeholders for different purposes than the farm. For example, by applying blockchain and using data provided by the different operators in the agro-food chain, the current high level of traceability can be enhanced to a level never witnessed by mankind. The increasing collection, use, and processing of data raise questions regarding the use of data, privacy, data protection, security, and ownership (Visvizi & Lytras, 2019).

Examples of Precision Farming Technologies

Today a farmer is able to locate its precise position in a field in a matter of centimetres, which allows the collection of real-time data with the help of various types of sensors (e.g., tractors or harvester with navigation system, sprayers, etc.) to create detailed maps of the spatial variability of as many variables as can be measured (e.g., crop yield, terrain features/topography, organic matter content, moisture levels, nitrogen levels, pH, etc.).

This information will be combined with data from multiple sources such as satellite imagery or drones, input supplier's data, meteorological data, in order to create the most accurate description of his crop enabling farmers to have a precise understanding of his crop.

Today, Galileo provides more accurate positioning and timing information for users enabling devices such as tractors navigation systems and mobile phones. Copernicus satellites support the timely and accurate monitoring of agricultural land use state.

Several agri-cooperatives are assisting their members in using new technological solutions (e.g., new machinery guided by GPS) to collect data from sensors on soil (pH, N), water, meteorological data, geodatabase, and satellites to combine with the experience of advisors and produce accurate reports that allow farmers to use the right dose in the right place at the right time.

Precision Livestock Farming

> Precision Livestock Farming (PLF) systems aim to offer a real time monitoring and managing system for the farmer. [...] The idea of PLF is to provide a real-time warning when something goes wrong so that immediate action can be taken by the farmer. This requires real-time algorithms that are able to detect or predict problems while the rearing process is ongoing. (Berckmans & Vandermeulen, 2013).

In livestock, different sensors on the farm (e.g., feeding and cleaning robots, milking robots, video analysis, etc.) allow to collect a significant amount of data (e.g., temperature, feed and water consumption, medicines, manure, animal behavior, etc.) that are combined in real time. With this information, the farmer gathers information concerning any potential animal health and welfare issue or the development of his animals, allowing him/her to adopt the most effective strategy to tackle any problem with the final goal of optimization of resources.

Forestry
In the forestry sector, real-time customer purchasing orders are combined with production data (collected by sensors, satellites, drones, etc.), which provide decision-making to the producers that, with the help of technology guided by GPS, make precise cuts (including lasers) of trees according to the specificities of the final costumer, saving time, transportation cost, and avoiding waste.

But Is It Only Primary Production Going Through Technological Transformation?

The uptake of technology and in particular the digital transformation allows improving the use of resources at farm level. In addition, it also helps to foster farming entrepreneurship and adapt business plans, as well as respond to dynamic markets and consumer expectations. It also optimizes the use of market-related tools, facilitate sales and competitiveness of EU farmers on a global scale, while respecting the environment and rural communities, decreasing administrative and bureaucratic costs, and enabling science-based policies. In addition, it also improves the relationships within the agri-food chain, allows for better marketing tools, simplification of administrative procedures, and contributes to provide better and more prosperous living conditions for rural communities.

One good example is the electronic wood marketplace[2] *puumarkkinat*. It is an electronic bulletin board where forest owners can market their timber lots

---

[2]Electronic wood marketplace www.puumarkkinat.fi

countrywide. The service is based on a map, where registered roundwood buyers have real-time information on private forest owners' roundwood supply. Roundwood buyers can evaluate, sort, and bid on lots, which are of their interest. The share of forest owners who live far away from their holdings has increased over the last decades. Since its inception, the value of marketed roundwood has exceeded €1 billion. In addition, at the end of September 2016, *puumarkkinat* had almost 1,100 users from forest management associations and roundwood buying companies. The next steps may allow producers to market these and other products worldwide without a heavy infrastructure.

The European Commission is also financing the 'The Internet of Food and Farm 2020' project aiming to consolidate Europe's leading position in the IoT-technology applied to the agri-food sector. The project wants to develop an ecosystem consisting of farmers, food companies, policymakers, technology providers, research institutes, and end users. The project aims to solve the European food and farming sectors' social challenges, maintain their competitiveness, and increase their sustainability. The IoF2020 project is organized around five agriculture sectors: arable crops, dairy, fruits, vegetables, and meat. Within each trial several use-cases (19 in total) show the value of IoT (Internet of Things).[3]

The relationship between the citizen and the public administration is also changing due to new technologies, in particular ICT and digital tools. Being a highly regulated sector, agriculture and food production will certainly benefit from the modernization of public administration services and the on-farm controls. In Estonia, all applications for public support have been digitized for some time leading to a reduction of the time to fill in the application from 3 hours (paper) to 45 minutes online. Each applicant has 24/7 access to his/her data files free of charge with a secure e-ID. The public administration has made available to each farmer interconnected public databases (e.g. soil maps, taxes, etc.) and his/her records.

The use of blockchain in case of food safety issues reduces reaction time from days to minutes (if not seconds), helping to save money and lives. It's possible to present and trace all information about the product from farm to fork. It can be used in the food sector so that each and every party along the length of the supply chain (producers, processors, and distributors) can provide traceability information about their particular role and for each batch (dates, places, farm buildings, distribution channels, potential treatments, etc.). It allows:

- full traceability of all sources of all inputs used in all stages in the chain;
- near real-time access to information on the Blockchain; and
- reduces costs of verification of attributes of products (e.g., product quality and prices) by eliminating intermediaries.

---

[3]https://www.iof2020.eu/

These are just some examples of the employment of technology in the agriculture and food production systems. Technology has been used to improve the current practices but soon technology will change the way agriculture and food production is actually done.

Future Trends of Technology

While the farming community is implementing some of the most advanced technologies, such as automated vehicles, blockchain, biotechnology, drones, and big data, substantial steps are being made in additional disruptive technologies that will have a huge impact not only in farming but also in the way we live, in particular in rural areas.

The data that the farming community is now collecting are starting to feed the algorithms driven by AI. AI provides the support decision systems that will be crucial to guide the automated robots that are aiming at performing many of the tasks that farmers can't find labor to do. These robots will work 24 hours a day, every day of the week. With IoT, they will be connected to everything. The European Commission has supported the Mobile Agricultural Robot Swarms (MARS) experiment aiming at the development of small and streamlined mobile agricultural robot units to fuel a paradigm shift in farming practices.[4]

Nanotechnology will allow the deployment of multitude of devices, including nanorobots, that will be able to monitor not only every leaf but every cell, increasing the quality, resolution, and quantity of data available. This will allow farmers to go from data-based decision-making to cell-by-cell decision-making.

3D will make possible the production of spare parts in the farm or making available substances such as veterinary or phyto-sanitarian products in a matter of seconds.

Why Is Precision Farming Important for Smart Villages Concept?

The first argument to be made is that while developing the key activity of most rural areas (agriculture and food production), PA and related services (e.g., machinery technical assistance or input suppliers) will provide the jobs and growth that will maintain a critical mass in rural areas. In fact, the author is convinced that the PA and the 'smart' food production processes will influence the design and implementation of smart villages.

The second argument is that the infrastructures and technologies needed to develop PF and 'smart' food production are also key to the development of the

---

[4]http://echord.eu/mars/

different services of smart villages (e-commerce, e-education, e-health, etc.) and vice-versa.

For example, the investment in broadband in rural areas is key to develop PA by allowing technological devices in farming to connect to not only data platforms, for example, but also to support the activities performed by an online shop of non-agricultural products or allow students to perform online research for school project or play videogames with their friends. In the same sense, the transport network will help the farmer to not only obtain inputs for his/her farming activity or to ship his/her products to other regions but will also be crucial for medical services/supplies to serve the population of smart villages (cf. Visvizi & Lytras, 2018).

An additional example is Galileo: the sentinels are key to support the navigation of the vehicles (e.g., drones, automated tractors, etc.). In addition, the same systems will also support automated ambulances to find the citizens in need for medical support or help a lost tourist to find the best route.

Technology employed in PA will be key to attract new talent into the farming business and provide the conditions to improve living conditions of farming families (remote work, access to online markets, etc.), ensuring that rural citizens may benefit from the same conditions as urban citizens.

But What Are the Bottlenecks?

Despite the potential benefits, the uptake of new technologies in rural areas and in agriculture and food production remains below what would be expected. In order to develop PA, a key component of smart villages concept, it is of utmost importance to tackle the different bottlenecks (connectivity, interoperability, skills, investment, R&D, etc.) that are slowing down the uptake of these technologies by the rural citizens and rural businesses.

A pillar for the uptake of ICT and digital tools is 'Infrastructure.' Today, Copernicus and Galileo, and the parallel services, are quickly being implemented but are still far from their full potential. This will allow the rural communities, agriculture, and food production to take advantage of vast amounts of data for free, including not only maps but also geo-referenced services that will allow machines to navigate with precision.

In order to take advantage of some of the services provided by Galileo for example, and to develop PF or other services related to rural life, it is absolutely critical to develop connectivity in rural areas. For example, the European Commission demands that all Basic Payment Scheme (BPS) applications be completed online by 2018. Unfortunately, not every farmer across EU has access to high-speed next-generation services. According to the last study of the European Commission on broadband coverage in Europe, although 92.4% of rural EU homes were passed by at least one fixed broadband technology in mid-2017, less than 50% (46.9%) had access to high-speed next-generation services.

Another key component for the development of PF involves the data processing capabilities that support farmer decision-making systems. Even though some work has been done in this field, a cross-sectorial and sectorial integrated decision-making system fit to farmers' needs is not yet fully available. These systems are key because they allow farmers and other agri-food businesses to make sense of all the data collected and transform it into valuable information that will create additional value.

Even if these decision-making support systems are in place, the collection and sharing of data raise many legal and ethical questions regarding privacy, ownership, use, and reuse of data. The European Commission has tried to tackle some of these issues by publishing the EU General Data Protection Regulation (GDPR) that replaces the Data Protection Directive 95/46/EC. The GDPR is a regulation on data protection and privacy for all individuals within the EU and the European Economic Area (EEA). It also addresses the export of personal data outside the EU and EEA areas.

When it comes to non-personal data in the agri-food chain, the farming community and the agri-food chain have published the EU code of conduct on agricultural data sharing by contractual arrangement. According to this voluntary initiative, farmers maintain the control and access to the data produced in their farm and during farm operations. The code defines key principles to respect when defining contracts for sharing non-personal agricultural data between agricultural businesses and their partners in the agro-food chain. These clarifications will facilitate data sharing, creating value and providing farmers and agro-cooperatives with clear and simple guidance. The goal is to improve transparency, responsibility, and trust when farmers and other operators share their data (CEMA, CEJA, EFFAB, FEFAC, CEETTAR, ESA, Fertilizers Europe, & ECPA, 2018).

The regulatory framework that fosters the adoption of innovative new technologies that are key to the development of PF has yet to be designed in certain key areas, such as new breeding techniques, Blockchain, or aerial spraying by precision drones. The regulatory framework is key to provide certainty about the use of these technologies, but it needs to be friendly to innovation, as otherwise they can actually hinder the technological development of these technologies or the uptake by the key operators.

Many other bottlenecks in areas such investment and financing, access to land, digital innovation hubs, and research and innovation need to be addressed in order to foster the uptake of PF and the services that are key for the smart villages concept. Nevertheless, there is not enough room in this chapter to address them all.

The author would like to conclude by underlining the need to set up integrated EU and national systems of upskilling and awareness. These must be complemented by relevant advisory and vocational training services. Skills, including digital skills, are an essential element of modern farm management. Providing the right training and education at the right moment will not only help farmers to make the most of these technological opportunities but will also produce educated digital entrepreneurs and an agricultural workforce that understands their rights and responsibilities in this new digital world.

Conclusions — What Are the Solutions/Way Forward?

Digital and technological transformation is changing the world economy, and digital business model patterns have now become relevant in physical industries as well. Farming processes are become increasingly automated, connected, and integrated, leading to the creation of models for data management and redefinition of the farmer–costumer–consumer relationship. These new technologies promise to increase the quantity and quality of agricultural output while using not only less input (water, energy, fertilizers, plant protection products, etc,) but also by connecting new and existent business models to the world economy.

Today, agriculture and food production are already investing and applying innovative solutions to keep their businesses competitive and sustainable, to better manage natural resources, to deliver public goods and services, to adapt to the various challenges, and to respond to societal demands. The technological and digital transformation of agriculture is no longer a discussion for the future but is rather the reality for many farms and agri-business all over the world.

The technological transformation of agriculture and the new data-supply chain places informed farmers in a new context and redefines their role in the supply chain, which will enable transformative agricultural business models to develop, leading to more transparency, as well as safer and better produce. Nevertheless, the farmer remains at the heart of all this transformation, and the agriculture and food production remains at the heart of smart villages.

In fact, many of the technologies, infrastructure, and skills being developed and implemented for farming will be used in other economic aspects and services in rural areas that will be key to successfully implement smart villages concept.

It is clear that agriculture is absolutely vital for a balanced development of European regions and the need to provide growth perspectives to rural villages of rural Europe. In order to reach this objective, it is clear that it is necessary to overcome the technological divide between rural and urban areas and to develop the potential offered by technology in rural areas.

Recommendations — How to Close the Gap?

One of the key features of *Homo sapiens* is his/her ability to understand 'time.' When planning a strategy, we understand the importance to think ahead, to look into the future. The best way to achieve meaningful change in the long term is to plan and execute a comprehensive strategy.

We have seen that technological transformation is vital for the future of agriculture and food production in the EU and to implement the smart villages concept. A comprehensive, coherent EU-wide strategy is then vital to help agri-food businesses and rural services to plan ahead and adopt the technologies that will fit their needs and feed successful business models.

It is the strategy, not technology, that is the key driver for not only technological transformation of agriculture and food production but also for the concept of smart villages that we want to implement. This strategy cannot be based on one-size-fits-all solutions but should be articulated with national and regional strategies adapted to the local conditions and specific needs. In this book we will not address the national and regional strategies due to clear lack of space, but the author will try to describe some of the key features of an EU strategy.

In order to implement the strategy, it's necessary to implement a set of policies and actions in order to reach the aim of the strategy. In the opinion of the author, policy is perhaps the most powerful tool ever created by mankind. The policies are the tools that make possible to accomplish the objectives that we as a society want to achieve.

The author believes that strong EU-level strategies and policies are key for thriving in the uncertain world of today and of the future. It provides the right scale for internal market to compete with other major economies and the flexibility to adapt to the big diversity of our agriculture and food production across the EU.

In order to implement smart villages concept described in this book, it is fundamental to have a thriving agriculture and food production that provides the jobs opportunities for the inhabitants along with the safest, affordable, and nutritious food that the latest technology allows. All EU policies and tools must align their efforts toward creating the right conditions to enable all farmers to adopt digital technologies and address potential challenges and minimize potential risks.

We therefore need a coherent strategy to promote a digital and technological transformation of EU agriculture and food production that enables *all* farmers and agro-businesses to connect. All EU policies must align their efforts toward creating the right conditions to allow all farmers and agro-businesses to uptake technologies that help farmers to adopt agricultural practices and technologies that are sustainable and at the same time maintain a competitive agro-business and raise rural incomes.

CAP is an obvious candidate to integrate this strategy. For more than 50 years it has successfully set the framework to provide affordable food for EU citizens while building the highest food safety, environment, health, and animal welfare standards. Given its agricultural orientation, the regulatory framework for farming activities and the financing that accompanies it is very well positioned to set the key objectives of the technology transformation strategy.

Research and innovation are among the tools that can provide the solutions fit to farmers' needs in order to tackle the many challenges ahead, including competitiveness and climate change, among others. Knowledge is key for the sustainability of the agriculture and food production sectors. The new research period wants to encourage increased investment in research and innovation in farming and food production. A specific budget of €10 billion from the Horizon Europe program will be set aside for research and innovation in food, agriculture, rural development, and the bioeconomy. The agricultural European Innovation Partnership (EIP-AGRI) will continue to pool funding sources from Horizon Europe and rural development to foster competitive and sustainable farming and forestry (European Commission, 2018c).

Today, the European Commission acknowledges that European citizens and businesses face many barriers to adopt digital technologies and their tools. In order to tackle these challenges, the Juncker Commission has proposed to create the digital single market:

> where the free movement of goods, persons, services, capital and data is guaranteed — and where citizens and businesses can seamlessly and fairly access online goods and services, whatever their nationality, and wherever they live. The single digital market is expected to tear down regulatory walls and moving from 28 national markets to a single one. This could contribute €415 billion per year to our economy and create hundreds of thousands of new jobs. A completed digital single market can help Europe hold its position as a world leader in the digital economy. (European Commission, 2018d)

In fact, the digital single market is a key tool for the successful implementation of PF but most important for the smart villages concept. From the objectives of the digital single market, there are three objectives that are key to the EU strategy of technological transformation:

(1) Unlock the potential of a *European data economy* with a framework for the free flow of non-personal data in the EU.
(2) Ensuring everyone in the EU has the best possible internet connection, so they can fully engage in the digital economy, the so-called "connectivity for a European gigabit society."

As we have seen earlier in this chapter, infrastructure is crucial for PF and smart villages (e.g., logistics, energy, health, and education). Access to high-speed broadband is key, but today there is still a considerable gap of broadband coverage between urban areas and rural areas.

> Rural broadband coverage continued to be lower than national coverage across EU Member States. Although 92.4% of rural EU homes were passed by at least one fixed broadband technology in mid-2017, less than 50% (46.9%) had access to high-speed next generation services. (European Commission, 2018e)

According to the European Commission, an increase of 10% in broadband connectivity in a country could result in an increase of 1% in GDP per capita per year. It also has the potential to increase labor productivity by 1.5% over the next five years.

(3) Helping large and small companies, researchers, citizens, and public authorities to make the most of new technologies by ensuring that everyone has the necessary *digital skills*, and by funding EU research in health and high-performance computing.

A strong digital economy is vital for innovation, growth, jobs, and European competitiveness. The spread of digital is having a massive impact on the labor market and the type of skills needed in the economy and society. On June 10, 2016, the European Commission published a new Skills Agenda for Europe, working together to strengthen human capital, employability, and competitiveness.

> It presents a number of actions and initiatives with the ambition to tackle the digital skills deficit in Europe. The new agenda sets out to improve the quality and relevance of skills formation, to make skills and qualifications more visible and comparable and advancing skills intelligence, documentation and informed career choices. (European Commission, 2018f)

Many other EU policies, such as trade, regional, and social policies, need to play a key role on the comprehensive EU strategy for digital transformation of EU agriculture and food production that will enable PF to show its full potential. These policies need to come together in one strategy in order to set up the baseline for not only business decisions in agriculture and food production, but also to provide the regulatory framework and incentives for private operators to provide the services (e.g., data processing capabilities) that match the farming community needs.

Agriculture and food production are the backbone of rural areas today. PF is step by step taking that role and will certainly be the cornerstone of the smart villages in the future. The author is convinced that it will also shape the design and implementation of this concept.

References

Berckmans, D., & Vandermeulen, J. (2013). *Precision livestock farming '13.* Papers presented at the 6th European Conference on Precision Livestock Farming, Leuven, Belgium.

CEMA, CEJA, EFFAB, FEFAC, CEETTAR, ESA, Fertilizers Europe, & ECPA. (2018). CEMA, CEJA, EFFAB, FEFAC, ..., ECPA. . *EU code of conduct on agricultural data sharing by contractual arrangement.* April 23, Brussels, Belgium. Retrieved from https://www.google.com/url?sa=t&rct=j&q=&esrc=s&source=web&cd=1&cad=rja&uact=8&ved=2ahUKEwiFkojRiq7fAhXQZ1AKHRiWBEEQFjAAegQICRAC&url=https%3A%2F%2Fcopa-cogeca.eu%2Fimg%2Fuser%2Ffiles%2FEU%2520CODE%2FEU_Code_2018_web_version.pdf&usg=AOvVaw2m-HLnofH0BvnGb2BBkIJQ

European Commission. (2015). *EU farms and farmers in 2013: An update.* EU Agricultural and Farm Economics Briefs No. 9, November. Retrieved from https://ec.europa.eu/agriculture/sites/agriculture/files/rural-area-economics/briefs/pdf/009_en.pdf

European Commission. (2017). *Communication on the future of food and farming.* COM(2017) 713 final, November 11. Retrieved from https://ec.europa.eu/agriculture/sites/agriculture/files/future-of-cap/future_of_food_and_farming_communication_en.pdf

European Commission. (2018a). *A clean planet for all — A European strategic long-term vision for a prosperous, modern, competitive and climate neutral economy*. Communication from the Commission to the European Parliament, the European Council, the Council, the European Economic and Social Committee, the Committee of the Regions and the European Investment Bank, November 28. Retrieved from https://ec.europa.eu/clima/sites/clima/files/docs/pages/com_2018_733_en.pdf

European Commission. (2018b). *A clean planet for all — A European long-term strategic vision for a prosperous, modern, competitive and climate neutral economy*. In-depth Analysis in Support of the Commission Communication (2018) 773, November 28. Retrieved from https://ec.europa.eu/clima/sites/clima/files/docs/pages/com_2018_733_analysis_in_support_en_0.pdf

European Commission. (2018c). *Horizon 2020 — Agriculture and forestry*. Retrieved from https://ec.europa.eu/programmes/horizon2020/en/area/agriculture-forestry. Accessed on December 20, 2018.

European Commission. (2018d). *Digital single market — Bringing down barriers to unlock online opportunities*. Retrieved from https://ec.europa.eu/commission/priorities/digital-single-market_en. Accessed on December 20, 2018.

European Commission. (2018e). *Study on broadband coverage in Europe 2017*. This study was carried out for the European Commission by IHS Markit Ltd., London, United Kingdom, June 2018. Retrieved from https://ec.europa.eu/digital-single-market/en/news/study-broadband-coverage-europe-2017. Accessed on December 20, 2018.

European Commission. (2018f). Digital skills & jobs. Retrieved from https://ec.europa.eu/digital-single-market/en/policies/digital-skills. Accessed on December 20, 2018.

European Union. (2018). Political declaration setting out the framework for the future relationship between the EU and UK. Retrieved from https://assets.publishing.service.gov.uk/government/uploads/system/uploads/attachment_data/file/759021/25_November_Political_Declaration_setting_out_the_framework_for_the_future_relationship_between_the_European_Union_and_the_United_Kingdom__.pdf. Accessed on December 28, 2018.

Lytras, M. D., Raghavan, V., & Damiani, E. (2017). Big data and data analytics research: From metaphors to value space for collective wisdom in human decision making and smart machines. *International Journal on Semantic Web and Information Systems*, *13*(1), 1–10.

Smit, H. (2015). *From intuitive to fact-based farming*. Food & Agriculture Research and Advisory, Rabobank Industry Note #513, Netherlands.

Visvizi, A., & Lytras, M. D. (2018). It's not a fad: Smart cities and smart villages research in European and global contexts. *Sustainability*, *2018*(10), 2727. doi:10.3390/su10082727

Visvizi, A., & Lytras, M. D. (Eds.). (2019). *Politics and technology in the post-truth era*. Bingley: Emerald Publishing.

WUR. (2018). *Precision agriculture — Wageningen University & Research, Smart Farming*. Retrieved from https://www.wur.nl/en/Dossiers/file/dossier-precision-agriculture.html. Accessed on December 20, 2018.

Chapter 7

Energy Diversification and Self-sustainable Smart Villages

James K. R. Watson

Introduction

The role of energy in smart villages is a central question for small communities looking to be self-sufficient, self-reliant, and digitally enabled. The objectives for communities vary around the world with smart villages in Europe usually seeking to find ways to maintain rural communities and a rural way of life, whereas in the developing world smart villages are often imagined as a means to escape poverty and deliver a sustainable future for the community. Energy is always a key concern in the development of a smart village concept and in this regard smart villages are being seen as a vanguard in the push for locally owned, decentralized energy systems both in the developed and developing world (Kammen, 2015).

This chapter will explore the experience of smart village projects in both Europe and across the world to compare the success that the projects have had in relation to utilizing clean energy to drive a sustainable smart village concept. Usually the key forms of energy that can be utilized by such communities are solar power, biomass, biogas, and solar thermal. The utilization of microgrids is also a common feature, and an enabling technology for the success of such smart village projects. These technologies are also suitable due to their inherent digital nature, particularly with regard to the use of solar power and microgrids which combines hi-tech elements that make the power supply even more efficient for village communities (Cheng, 2014).

Given the inherent decentralized and digital potential of solar technology, much of this chapter will focus on the utilization of solar power for smart village communities. Its versatility is well recognized and its ability to be installed on roofs, by roads, on sign posts, walls, and barns makes it a natural ally of the smart village concept (Gent, 2016).[1] It is also easy to use and has very few toxins for end-of-life concerns, being broadly an inert technology (IRENA, 2016). Solar has also experienced a massive reduction in cost and a large increase in efficiency in just the past decade, which again makes it a natural solution for many rural communities. This two-fold increase in the benefits of solar technology may help many communities across the globe achieve their objective of having power in a self-sufficient manner.

In the future, the advent of batteries may also contribute to the self-sufficiency of smart villages in parts of the world, and much depends on the sustainable production and end-of-life systems that are introduced. Technically, batteries are a perfect partner for solar and microgrids and can keep the power going on through the night or even days at a time. Indeed, some small Pacific islands already use solar and batteries with a microgrid to power their communities (SMA, 2017). In communities with more biomass that is sustainably available — particularly waste from plants and animals — biogas digestors will also become an increasing source of 24/7 power for communities. It is likely that every smart village concept will identify the type of sustainable power generation that works best for it, and this will vary considerably across regions and countries.

This chapter will explore the role that energy plays in driving smart villages, first looking at solar energy, which has been widely acknowledged as an opportunity for powering smart villages. The other suitable technologies will also be examined, particularly biogas, to demonstrate the wide options that exist for sustainably powering smart villages. The chapter will then consider the role the energy plays in driving digitalization in rural communities, before concluding with an assessment of the role that clean energy can play in driving the smart village phenomenon.

Solar-powered Smart Villages

Solar power has been a catalyst for many smart village projects across the world, particularly in Africa, India, and Latin America. Solar power has also been utilized in Europe by village communities seeking a clean and reliable source of energy. A quick glance across the smart village literature shows that in terms of energy self-sufficiency and clean sourced energy, solar energy is one of the key drivers for communities looking to bring development and services into their communities (Kumar & Shekhar, 2015). Indeed, many projects have based their whole smart village concept on the basis of solar power delivering the vital means of energy to deliver the overall infrastructure of the project (EkoEnergy, 2018). This is a trend that is in fact growing and can be expected to grow as countries seek to bring energy to their rural communities in a cost-effective manner, yet in a clean manner to ensure that they meet the objectives of the Sustainable Development Goals (SDGs) and their Paris Agreement commitments.

---

[1] See ref. Gent (2016).

The first question to address is why has solar technology been so successful in supporting smart village projects and communities. Why have communities and governments turned to solar energy to drive the development of smart rural communities? The answer to the question can be given in many parts. First, to achieve the SDG7, governments need to provide people the means and opportunity to have reliable, clean, modern, and sustainable energy (UN, 2015), and the Paris Agreement commitments on climate change require countries to cut their carbon dioxide emissions (UNFCCC, 2015). Thus, to achieve SDG7 and taking into account the Paris Agreement, the need for investment in clean technologies is clear. Solar technology is part of the renewable energy technologies that can provide clean, modern, zero carbon dioxide emissions and modern energy to communities. Therefore, it clearly fits the bill for governments and thus communities to play their part in delivering both the SDGs and the Paris accord.

Solar energy has also seen a considerable price reduction in the past 10 years. In fact, it is arguable that any other technology has seen as quick a cost decline as solar energy has achieved in the past decade. Since 2010, solar power prices have fallen by 70% according to the International Energy Agency (IEA), and the trend is expected to continue (IEA, 2018). This almost unprecedented cost reduction in renewable energy is standing solar power in good stead to become the world's leading means for power generation by 2050. This cost reduction also makes solar power a very attractive option for smart villages who generally seek cost-effective means to deliver on their concept. The fact that it is also increasing in efficiency means that communities can combine reduced costs with increased power output, the old adage 'more bang for your buck' has never been truer for solar. Efficiencies have grown from about 12% in 2010 to around 20% for super-mono cells (Philipps, 2018). This means an increase in yield that has almost doubled – at the same time the cost has fallen by 70%. Finally, the technology also offers an increasing durability. Today it is estimated that a solar panel will still produce around 92% of the initial power output after 20 years (Jordan & Kurtz, 2012). This is up from estimates of around 80% for modules produced prior to 2000 (Jordan & Kurtz, 2012). As the technology improves and further matures, this will likely increase further in the coming years. Altogether these points of cost, production, and longevity make solar a perfect technology to drive smart villages.

Other important facets of solar technology make it ideal for the utilization of smart villages, not least its incredible versatility. As mentioned earlier in the chapter, the use of solar power can be almost anywhere that can safely support the structure. A panel weighs around 10 kgs per square meter and can therefore easily be supported by most roof structures around the world (Aaron, 2016). In any case, a panel can be placed against a south-facing wall where roof structures are not suitable. There are many ways to get the most out of a panel for villagers around the world. In Europe, panels are also being used in roads for some communities (Murphy, 2016). Thus, panels can be placed anywhere and make good quantities of clean energy for smart villages. Indeed, in the world today, almost 40% of all the 400 GW of solar installed is on rooftops (or equivalent) (IEA, 2018). In the European Union (EU) this share is as high as 70% of rooftop solar panels as compared to utility scale (Schmela, 2018). Thus, the rooftop potential of solar technology means that it is indeed easy to integrate into smart village concepts as it is truly decentralized. This is a characteristic that is not shared by many other renewable technologies, except for biogas and solar thermal for hot water.

Added to these technological advantages, it is also relatively easy to connect to microgrids, and due to the digital nature of the solar system's inverter, it is also highly controllable from a person's smartphone. Thus, there is a highly personal nature to the utilization of solar power which sits well with smart village community projects. The digital nature of solar power also makes it a technology that is utilizable in many countries that have experienced a strong uptake in mobile phone usage, as simple apps can be used to control output and frequency. In the United States, latest projections show that solar is going to become the dominant technology attached to its microgrids in the near future, as planned additions in solar technology outstrip traditional technologies like combined heat and power (CHP) and diesel (Thurston, 2018). In the rest of the world, there is also potential; in India, many solar microgrid initiatives have been set up to try and kick-start smart village communities; however, not all have been successful as there are many non-technology issues to overcome such as promises of grid extensions from politicians, which may never be realized but offer electricity much more cheaply, and electricity theft, which is a major concern (Fowlie, 2018). Nevertheless, it has been shown that in India consumers are more satisfied with their solar microgrids than those customers who utilize the normal grid, which suggests a strong reason for optimism in the continued successful roll out of solar and microgrids (Graber, Narayanan, Alfaro, & Palit, 2018). Solar energy is also highly suitable for modern communication technologies, which can make a win-win for the smart village. Charging your mobile phone with a solar panel is a convenient and cheap way for smart villagers to maintain their communication channels, as well as control their power source (Kammen, 2015).

These characteristics suggest that solar power will be the key technology to drive smart villages across the world, bringing work, services, healthcare, education, communications, and infrastructure to isolated − yet increasingly connected − communities.

Other Power Technologies and Smart Villages

Clearly solar and solar-powered microgrids are going to be at the forefront of smart villages across the world. The preceding section lays the arguments for this. However, solar technology is not the only technology that will be available and utilized by the inhabitants of smart villages to bring clean and cheap power to them. Other forms of power generation will also be available and utilized in smart villages. In India, there have already been some successful projects where villages utilized biogas as a means to provide power and support their cooking activities (Nasery & Rao, 2011). In India, the potential for biogas is alone estimated at 19,500 MW, only a little less than solar, and is derived from the estimation of agricultural residues — both plant and animal that are available (Nasery & Rao, 2011). This makes biogas particularly useful in the Indian situation and particularly for smart villages, as it can provide many uses that substitute environment harming activities. For example, biogas can be used instead of firewood, reducing local deforestation, as well as diesel, or by simply burning agricultural residues. Biogas can derive energy for cooking and lighting, as well as provide fertilizer through the residues formed from the anaerobic digestion that takes place in the process of producing the biogas. Another important element is that biogas can be fed by human waste as well as animal waste and thus can offer a very useful sanitation and waste disposal system for rural communities and support the creation of smart villages (Nasery & Rao, 2011). Inadvertently this can also help stop the spread of pathogens.

Biogas can also power rural industries, thus creating employment, and can also be bottled and sold as another means of generating income in rural areas (Templeton & Bond, 2011). Thus, biogas can also offer a set of economic activities that can present opportunities for smart village communities. In this regard, it is not surprising to see that India has embraced this technology, and it was estimated in 2007 that there were already 2.5 million household and community biogas plants installed in India (Dutta, Shields, Edwards, & Smith, 2007). Biogas is seen as a source of energy that matches well the idea of smart, devolved, self-sufficient villages. Combined with solar power, it can also provide energy around the clock at a cost-efficient level, with little need to be concerned about the end of life of the apparatus, such as is associated with other technologies at the end-of-life stage.

In Germany in the rural region of Göttingen, a Bioenergy Villages project has been set up that aims to improve the renewable energy provision for its 120 villages (European Network for Rural Development, 2018). The initial interest was apparent with 34 villages applying, and already five have achieved the status of operational Bioenergy Villages. There is very strong citizen participation in the process, and this creates ownership and buy-in to the objectives of the projects. All the projects are based on biogas plants, with some wood chip furnaces and local microgrids. Farmers and village cooperatives work together for energy production and distribution. This provides the farmers an extra income stream as they can provide waste and agricultural byproducts for the purpose of energy generation. The financial benefit is just one of a handful of benefits that can be seen from the projects: the villages also gain new skills as they trained in using new technologies, the villagers learn how to apply for funding for projects — a skill that can be reused for other smart initiatives — and in turn they can also build village knowledge centers and multiservice hubs that provide information and wealth generation (European Network for Rural Development, 2018).

The use of biogas is a growing trend to develop smart village projects, but the use of biomass has in general always been a part of most rural lives across the globe.

The use of biomass can continue to support smart villages if it is harvested and used in a sustainable fashion. Traditionally for many rural communities wherein biomass has been burnt for cooking and heating, it has usually come in the form of wood or other plants and plant waste. The upside for the communities has been the relative abundance of plants to burn for energy; however, this is becoming less reliable as desertification and deforestation, often as a result of climate change, take their toll on rural communities. The advent of smart villages can seek to address the problem of reducing biomass availability for energy by offering alternatives. Nevertheless, in a step-wise approach it is likely that biomass will continue to be used – even in a hybrid fashion alongside solar power and batteries (Ho, Hashim, & Lim, 2014). The downside has been related to the loss of forests and habitats but has also taken a toll on health as biomass burning for cooking creates soot and smoke that impact the residents' lungs and overall health. Therefore, moving to hybridization and then on to solar power or biogas also offers opportunities for communities to move forward step by step to improving their lives (Ho et al., 2014).

Other technologies such as wind and solar thermal will also play a role in the provision of energy for smart villages, but this segment has focused on the most promising alternative to solar – biogas. However, the combination of biogas and solar power could be ideal in many developing countries to provide round-the-clock clean energy and simultaneously improve the health of rural communities as well as providing opportunities for new skills to be developed by the inhabitants.

Digitalisation and Energy in Smart Villages

Smart villages of course require modern, clean, renewable energies to deliver a better future for their inhabitants, but a village is not smart just because it has clean energy. It is also important that the villages are connected and have digital capabilities. Indeed, to maximize the potential of a technology like solar power and a microgrid it is necessary that there is sufficient digital access for the inhabitants. The need for high-speed internet and wifi access is as important as the provision of the power itself, for the long-term sustainability of the project. In this regard, the use of smart energy can also support and drive the digitalization of the village communities and increase opportunities not just from the benefits of clean energy but also from increased access to sell produce and gain access to the market of the internet.

One of the main drawbacks in the EU has been the lack of investment of governments in the development of rural high-speed internet links. Indeed it is noted that only 25% of rural communities in the EU have high-speed internet, compared to 70% of their urban counterparts (Stam, 2018). Bringing information technology (IT) services to rural communities is also a key component of the drive for smart villages and sustainable villages. It is important that smart villages can provide jobs and services for their people, but they must also deliver on connecting the community to the rest of the world (Government of Rwanda, 2015). This may support local development and reduce the likelihood of brain drain in rural areas that so badly affects the rural populations across the world.

IT skills are also closely linked to the development of clean energy technologies such as solar energy. Solar energy is a digital technology and can be controlled remotely through the functioning of its inverter. The power production of solar panels can also be monitored digitally so that it is possible to see the production per cell and also identify areas of underperformance (SolarPower Europe, 2018). This digital nature of solar panels means that operating and maintaining solar power offer opportunities for training in more digital skills, which can support the development of jobs and a more resilient workforce in rural areas. Solar energy also offers the opportunity for development of blockchain systems in small areas – where peers can sell electricity to one another. However, for this to work well, the need for digital services in the form of high-speed internet become increasingly important as villagers seek to maximize the utilization of their solar systems (SolarPower Europe, 2017). In this regard, as more solar power is installed in villages, there is a need for more digitalization of the community, and thus the use of this versatile clean energy can result in a drive for improved internet connection.

It is therefore the case that it can be argued that solarisation of smart villages will also drive the need for communities to be serviced with high-quality internet connections. This really is a win-win scenario. This is particularly true if the deployment of solar power is accompanied by the development of smart grids. It has been noted that where microgrids have been utilized they rely as much on information systems as they do on direct energy technology (Kammen, 2015). Microgrids and home solar systems have benefitted from becoming much cheaper, as we have seen, and developments in their performance – which also relate to mobile communications technologies that can also increase access to internet banking, which really can benefit the economy and standard of life of villagers (Kammen, 2015). This heady mix of innovation delivers new services that suggest off-grid smart villages can become strong economic centers for rural communities.

The Outlook for Clean Energy Driven Smart Villages

The EU has set up an initiative to increase the number of smart villages in the EU, and with the strong commitment of the EU to achieve the Paris commitments that they have undertaken, clean energy will be a core element of the development of smart villages (European Network for Rural Development, 2018). In the recently concluded 'Clean Energy Package,' the EU agreed to introduce a right for small consumers to self-consume their energy without hindrance in terms of taxes, tariffs, or charges. It also empowered local energy communities to be developed under their own direction. These developments suggest that the EU will seek to push further into the realm of smart villages, as indeed specific funding has been earmarked for the development of smart villages in the forthcoming seven-year budget of the EU (Stam, 2018). With a protective legal framework in place for small-scale energy production, such as that provided by solar and biogas, it is likely that the clean energy element of smart villages will be a driving factor in the increased uptake of smart village projects. The EU's smart village funding project is being targeted to support at least 10 villages across the EU (European Network for Rural Development, 2018). Coupled with national EU initiatives, the outlook for the development of rural communities and decentralized energy solutions is bright.

In other regions of the world many governments working with international partners are also actively bringing forward smart village concepts, often with clean energy solutions at the forefront. India has been cited many times in this chapter and has indeed had its own version of the smart village initiative since 2014, when Prime Minister Modi launched his Sansad Adarsh Gram Yojana programme (Government of India, 2017). The program does not have energy as a key factor per se, and has faltered in recent years due to lack of pick up by Members of Parliament (MPs) in India, but does provide the opportunity for rural communities to approach the issue of how digital and energy needs can be met through political intervention. Many other programmes at regional and state level are also ongoing in India with varying rates of success, and in general, it seems that India will remain a good prospect for developing smart villages now and in the future.

In Africa, different countries have introduced different programmes to support villages; however, many fall into the category of being considered as smart village initiatives. In Rwanda, the government has worked with the United Nations Development Program (UNDP) and United Nations Environment Program (UNEP) to create a 'Toolkit for the development of Smart Green Villages' (Government of Rwanda, 2015). The initiative couples the need for digitalization of communications in villages with the provision of clean energy to improve the villagers' lives. In this case, the focus is strongly on the provision of solar-powered lamps to replace kerosene ones, which provide solar power even after dark thanks to the use of rechargeable batteries that store the power gained during the day. The project also focuses on the development of biogas for cooking, thereby again reducing the dependency of the communities on biomass and oil. Introducing clean energy and providing IT support and communications infrastructure also form part of the program. The program was piloted in two villages, and the success of the program has encouraged the Rwandan government to role out the program across the country. Thus, in Rwanda, and across much of Africa where similar experiences are occurring, it is clear that the smart village concept has a very strong future.

This shows that an integrated approach to rural development and the challenges facing village communities can result in improved lives and opportunities across all the regions of the world. One can only conclude that the advent of smart villages will begin a new age of development in rural communities, empowering people to take control of their lives in terms of energy and communications for many years to come.

Conclusion

This chapter has explored the opportunities for smart villages in terms of clean energy and has examined the mutuality between renewable energy sources and IT infrastructure development for rural communities. It is very clear that smart village projects benefit from embracing new energy technologies like solar and biogas, as they empower villagers and also bring an element of smart new technology to the communities. These two elements alone are major benefits, but smart energy also brings employment, health benefits, skills, and services that would otherwise not be present in such communities. Therefore, it is clear that the role of clean energy is vital in achieving truly smart villages.

The fact that solar energy in particular can drive change and improved communications infrastructure for rural communities also indicates a strong likelihood that in the future solar power will be a key element of smart villages across the world. Biogas will also be a key energy technology for rural communities seeking a way forward from traditional energy sources and fossil fuels and will increasingly appear across the globe. Other technologies will also be utilized, but the focus is likely to be on these two champions of decentralized energy.

In the future, smart village projects are likely to spring up across the globe, judging by the success of the programs that exist in Europe, Africa, and India today. The only question is how fast these programs can deliver and whether governments will be able to effectively address challenges as they arise, such as electricity theft. In all likelihood, the smart village driven by clean energy will become the normal village of the next decade.

References

Aaron, K. (2016). How heavy are solar panels? Retrieved from https://www.renewenergy.com.au/how-heavy-are-solar-panels/

Cheng, G. (2014). A solar power microgrid. *North American Clean Tech, 8*(4).

Dutta, K., Shields, K. N., Edwards, R., & Smith, K. R. (2007). Impact of improved biomass cookstoves on indoor air quality near Pune, India. *Energy for Sustainable Development, 11*, 19–32.

EkoEnergy. (2018). Solar powered smart village. Barbujat, Sudan. Retrieved from https://practicalaction.org/barbujat

European Network for Rural Development. (2018, May). Barriers to community led innovation. EU Rural Review No. 26. Retrieved from https://enrd.ec.europa.eu/sites/enrd/files/enrd_publications/publi-enrd-rr-26-2018-en.pdf

Fowlie, M. (2018). Are solar microgrids the future in the developing world. Energy Institute Blog. Retrieved from https://energyathaas.wordpress.com/2018/10/01/are-solar-microgrids-the-future-in-the-developing-world/

Gent, E. (2016). How India's 'smart villages' are centalising solar power. BBC website. Retrieved from https://www.bbc.com/news/world-asia-india-36681112

Government of India. (2017). Sansad Adarsh Gram Yojana Programme. Retrieved from https://en.wikipedia.org/wiki/Smart_Village_India

Government of Rwanda. (2015, June). A toolkit for the development of smart green villages in Rwanda, Rwanda Environment Management Authority, UNDP & UNEP.

Graber, S., Narayanan, T., Alfaro, J., & Palit, D. (2018, February). Solar microgrids in rural India: Consumers' willingness to pay for attributes of electricity. *Energy for Sustainable Development, 42*, 32–43.

Ho, W. S., Hashim, H., & Lim, J. S. (2014, September). Integrated biomass and solar town concept for a smart eco-village in Iskandar Malaysia. *Renewable Energy*, *69*, 190−201.

IEA. (2018). Solar PV tracking clean energy progress. Retrieved from https://www.iea.org/tcep/power/renewables/solar/

IRENA. (2016). End-of-life management solar photovoltaic panels. St. Ursen, Switzerland: IEA PVPS and IRENA.

Jordan, D. C., & Kurtz, S. R. (2012). Photovoltaic degradation rates − An analytical review. *Progress in Photovoltaics Research and Applications*, *21*(1), 1−11.

Kammen, D. M. (2015). Energy innovation for smart villages. In B. Heap (Ed.) S*mart villages: New thinking for off-grid communities worldwide*. Cambridge: University of Cambridge.

Kumar, D. Dr., & Shekhar, S. (2015). Rural energy supply models for smart villages. In M. V. Rao, K. P. Rao, P. T. Kumar, & D. S. R. Murthy (Eds.), *Geoinformatics applications in rural development* (pp. 166−178). Hyderabad: Professional Books Publisher.

Murphy, M. (2016). The world's first solar panel-paved road has opened in France, Quartz. Retrieved from https://qz.com/871162/the-first-road-paved-in-solar-panels-opened-in-france/

Nasery, V., & Rao, A. B. (2011). Biogas for rural communities. Centre for Technology Alternatives for Rural Areas, India Institute of Technology Bombay. Retrieved from https://www.cse.iitb.ac.in/~sohoni/pastTDSL/BiogasOptions.pdf

Philipps, S. (2017, July 12). Photovoltaics report. Freiburg, Germany: Fraunhofer Institute for Solar Energy Systems.

Schmela, M. (2018). Global market outlook for solar power. SolarPower Europe. Retrieved from solarpowereurope.org/wp-content/uploads/2018/09/global-market-outlook-2018-2022.pdf

SMA. (2017). Retrieved from https://www.sma-sunny.com/en/st-eustatius-100-solar-power-in-the-caribbean/

SolarPower Europe. (2017). Solar digitalisation report. Brussels, Belgium: SolarPower Europe.

SolarPower Europe. (2018). Operations and maintenance guidelines. Brussels, Belgium: SolarPower Europe.

Stam, C. (2018). Smart villages are an opportunity to bring jobs, says MEP Bogovic. *Euractiv*, April 13. Retrieved from https://www.euractiv.com/section/agriculture-food/news/smart-villages-are-an-opportunity-to-bring-jobs-says-mep-bogovic/

Templeton, M. R., & Bond, T. (2011, December). History and future of domestic biogas plants in the developing world. *Energy for Sustainable Development*, *15*, 347−354.

Thurston, C. W. (2018). Solar emerges as top new microgrid energy source. Clean Technica. Retrieved from https://cleantechnica.com/2018/10/26/solar-emerges-as-top-new-microgrid-energy-source/

UNFCCC. (2015). Paris Agreement. Retrieved from https://unfccc.int/files/meetings/paris_nov_2015/application/pdf/paris_agreement_english_.pdf

United Nations. (2015). Sustainable Development Goals, Number 7. Retrieved from https://sustainabledevelopment.un.org/sdg7

Visvizi, A., & Lytras, M. (2018a). Rescaling and refocusing smart cities research: From mega cities to smart villages. *Journal of Science and Technology Policy Management(JSTPM)*, doi:10.1108/JSTPM-02-2018-0020

Visvizi, A., & Lytras, M. D. (2018b). It's not a fad: Smart cities and smart villages research in European and global contexts. *Sustainability*, *2018*(10), 2727, doi:10.3390/su10082727

Wolski, O. (2018a). The place of rural areas in regional development concepts and processes. In P. Nijkamp & K. Kourtit (Eds.), *Rurality in an urbanized world*. Maastricht: Shaker.

Wolski, O. (2018b). Smart villages in the EU policy: How to match innovativeness and pragmatism? *Village and Agriculture*, *4*, 181.

Zavratnik, V., Kos, A., & Stojmenova Duh, E. (2018). Smart villages: Comprehensive review of initiatives and practices. *Sustainability*, *10*, 2559. Retrieved from https://www.mdpi.com/2071-1050/10/7/2559

Chapter 8

The Role of Smart and Medium-sized Enterprises in the Smart Villages Concept

Xénia Szanyi-Gyenes

Introduction

When the European Commission talking about smart villages, it is understood that they are with a population below 10,000 inhabitants; however, it should not be necessarily a single settlement, but a network of connected villages is also the subject of the matter. Smart villages may develop the rural areas and SMEs should play a significant role. Small- and middle-sized enterprises (SMEs) are crucial to Europe's economy, anyway. There are more than 20 million SMEs that play a key role in the economic growth of Europe, as well as in the innovation sector and in the field of job creation. Smart villages are rural areas and communities, which build on their existing strengths and assets as well as on developing new opportunities, where traditional and new networks and services are enhanced by means of digital, telecommunication technologies, innovations and the better use of knowledge (European Commission, 2018a).

In smart villages, traditional and new networks and services are enhanced by means of digital, telecommunication technologies, innovations and better use of knowledge, for the benefit of inhabitants and business (European Commission, 2017). In the literature, there is a special definition of the smart village:

> (i) A village is an ecosystem of limited size, a community that is driven by specific mechanism and dynamics that are the product and the outcome of multi-level interaction among all stakeholders; (ii) smart village is conceptually different that the aggregate construct of a "rural area" or "country side"; (iii) a village is conceptually and empirically distinct, and so the question of and corresponding research on the value-added [information and communication technology (ICT)] can garner in the space of a village has its own characteristics, different from research on smart cities. (Visvizi & Lytras, 2018b)

In the same debate, it is also argued that a cautious rethink of the focus and the rationale behind the smart villages debate is needed to avoid the 'ICT-hype', so specific to the smart cities debate and ensure that ICT-enhanced solutions are useful and usable by rural areas inhabitants (Visvizi & Lytras, 2018b). As the debate on smart villages is taking only its shape, and considerable effort needs to be invested in delineating its conceptual and empirical boundaries (Visvizi & Lytras, 2018a), this chapter does so by focusing on, otherwise absent in the discussion, the connection between SMEs and smart villages. Accordingly, the objective of this chapter is to give an overall picture about the possible connection between the SMEs and smart villages. A systemic approach is needed, where the significant problem could arise from the harmonization of the system, at three levels, including: (1) at the local level; (2) between the local levels; and (3) between the nations.

Based on current experience there is no adequate coordination between a lot of possibilities of SMEs and there is no ring-fencing and dedicated allocation of budget for smart villages on funds' level. Therefore, a smart coordinated system is necessary. The elements of the system should work complementing each other and cooperating with each other.

Acknowledging the opportunities for these SMEs, this chapter seeks to answer the following central research questions: How can we actively involve SMEs in the strategic planning and the implementation of these projects? What can be considered as local smart solution?

The basis of this chapter is a short literature review, including recent studies on the effectiveness of SMEs in the EU. It keeps in focus the need for rethinking the cooperation between villages and SMEs for the sake of successful smart villages. Besides that, some smart examples are presented to show the possible synergies in practice. The examples are used to reveal some specific features of new trends and tools for smart villages, including the role of networking and cooperation, fully exploiting the benefits of smart networks.

The argument is structured as follows: The chapter proceeds with a discussion of the future solutions for SMEs and concludes that the cornerstone and the engine of smart villages is the SME sector with an appropriate certificate system. While a more in-depth research is needed to confirm these results, the chapter directs attention toward this sector. It states that with the help of stakeholders, we can enable the basis for smart villages to manage the circularity of economies in the given villages.

The basic hypothesis of the chapter is that SMEs could play a key role, but their potential is not fully exploited in the smart villages concept. Rural SMEs may concentrate on local needs. They have the capacity to find local solutions for local problems. There are many valuable building blocks for smart villages, among them the small companies. But more than building blocks, we need to make synergies too. We need to create new business models, which are based on connection, integration, and cooperation.

The Role of SMEs in the European Economy

Small- and medium-sized enterprises, but rather micro-businesses, can be called the spine of European economies. They have a central role in growth, employment, and innovation. According to the 2014 Annual report on European SMEs, there are nearly 22 million small- and medium-sized enterprises operating in the EU. At the same time, it is a very important statistical indicator that SMEs account for almost 99% of EU businesses in number, according to the EU SME classification. In the EU, these SMEs account for almost two-thirds of the private sector and 58% of the gross-added value generated by businesses. That is why support for SMEs is a major priority. According to the EU SME classification, we distinguish three categories: microenterprise, small business, and medium-sized enterprises. The rating is based on the number of employees, the turnover, or balance sheet total (Szanyi-Gyenes, Mudri, & Bakosné, 2015)

Microenterprises are defined as enterprises which employ fewer than 10 persons and whose annual turnover or annual balance sheet total does not exceed two million euros. Microenterprises represent the biggest group: more than 90 percent of SMEs. This group includes mainly private companies and family businesses. Small enterprises are defined as enterprises which employ fewer than 50 persons and whose annual turnover or annual balance sheet total does not exceed 10 million euros. Medium-sized enterprises are defined as enterprises which employ fewer than 250 employees and whose annual turnover does not exceed 50 million euros and annual balance sheet total does not exceed 43 million euros (European Commission, 2018b) If we use this classification, the majority of the SMEs in the EU were small and microenterprises in 2013.

Small companies play a significant role in the economy, but even more in the field of job creation and innovation. Innovation has outmost importance nowadays, while the environmental aspects have to be taken into account for our viable future. SMEs represent over 90% of all EU businesses and account for two of tree jobs (in 2013, 21.6 million SMEs employed 88.8 million people); therefore, we can state that SMEs are the backbone of the European economy, and we can say that they could assist in the success of smart villages (European Commission, 2015a). Looking at the statistics, about 80 percent of all SMEs operate in five key sectors. The most important SME sectors are wholesale and retail trade sector, and the largest SME sector in the EU: manufacturing; construction; professional, scientific and technical activities; accommodation; and food. Together, these five sectors account for almost four-fifth of all SMEs in the EU-28. Some SME sectors has posted relative strong positive growth from 2008 to 2013 with "business services", "retail and wholesale trade" and "other sectors" – which included all other non-financial business sectors – posting positive value-added growth (European Commission, 2015b). These sectors may play a key role in developing services for the smart villages. SMEs are at the forefront of innovation, and they are able to understand and serve the local needs. This may be the basic to create an operative system of smart villages. There are more than 20 million SMEs in the EU, and 90% of them are microenterprises. Smart villages need the participation of these small companies.

That is the statistical fact and that is why it is in the center of the political and research interest. Because of the crucial role of small companies in the economy, it is important to examine them from as many aspects as possible in order to understand their motivation and the essence of their operation. Furthermore, it can give a compass to the economic and political leaders so that they know what these companies need, and therefore, they can create effective support resources.

The Role of SMEs in Innovation

When we examine the importance of small business, we should talk about SME's role in innovation. Europe's entrepreneurs have always been at the cutting edge of innovation and invention. And this is true even more today. The EU actively supports SMEs by providing both direct financial support and indirect support to increase their innovation capacity. This is well-reflected in the EU's proposed budget outline for 2021–2027 in which a new instrument would allocate 10 billion euros for innovation in the agri-food sector with a clear focus on small businesses. We should therefore assist the creation of a favorable ecosystem for SME innovation and growth. The objective is to optimize the research, development, and innovation environment for SMEs, through the establishment and facilitation of a range of support services, with the aim of strengthening the innovation capacity of SMEs and creating value on the market and/or into society (European Commission, 2018c).

Exploiting this innovation's potential has become more challenging than ever, in particular, for many SMEs. The globalization, the high degree of interconnectivity, and new technological advances have forced many SMEs to reinvent not only their products and services, but also their business and organizational models, in view of staying competitive in a fast-paced economy (European Commission, 2018d).

Indeed, EU policies often focus on encouraging and stimulating innovation. Its role is creating job opportunities, increasing the competitiveness of enterprises in global markets, improving the quality of life and contributing to more sustainable economic growth. Almost half of all the enterprises in the EU-28 reported some form of innovation activity (49.1%) during the period 2012–2014. Among the EU Member States, the highest proportions of innovative enterprises during this period were observed in Germany (67.0% of all enterprises), Luxembourg (65.1%) and Belgium (64.2%), while Ireland and the United Kingdom also recorded proportions that were above 60.0%. (Eurostat, 2017), where an innovation-active firm is one that has had innovation activities during the period under review (Eurostat, 2013a) and innovation activities are all scientific, technological, organizational, financial and commercial steps which actually, or are intended to lead to the implementation of innovations. (Eurostat, 2013b) It is difficult to analyze these data since EU's innovation statistics of 2017 relates to that of the 2012–2014 period, and planned article update is expected by January 2019.

Small Companies and Sustainability

The research is based on the assumption that the role of small businesses in the economy is dominant. Furthermore, the role of small businesses in sustainability is also increasing. For purely political reasons, governments prefer to lay down rules that are tailored to acquire certain potential or real giant investments, which obviously do not serve the special development needs of the SME sector. Such distortions include, for instance, more difficult access to support too many required references and a number of concessions whose interpretation is almost impossible for small businesses. The favoritism toward multinational companies is based on the assumption that with their appearance they would boost the supplying industry for the local businesses. It is partially true, but it would be relevant to the whole of the economy only if the multinational companies passed on the benefits they had received to their smaller partners, including tax exemptions. Note that this issue is further exacerbated by the fact that the EU competition law compliance investigation and proceedings almost always came to the conclusion that the big corporations practically do not pay their taxes – Google, Apple, etc.

Usually, small businesses can apply for generally applicable resources and not for a dedicated purpose. The most important step would be to set up a funding framework specifically designed to develop solutions for smart villages. The rules should be simple and easy to follow. Small business can have great difficulties with administrative burdens. Their tenders need to be coordinated and unified at a national level, that is how it can become an integrated and successful European development project.

Smart Solutions for SMEs

Yet, in the smart villages concept, we need a new mechanism. It is important to create a smart environment in which entrepreneurs and family businesses can thrive and build their own smart villages. We need to create rules in line with three basic principles: (1) find real local solutions; (2) sustainability; and (3) efficiency. Administrative systems should meet the needs of SMEs and smart villages. We need to adopt the policy toolkit to SMEs and facilitate the access to finance for SMEs. SMEs need to be helped; so they can enjoy the benefits of the rural areas. We should promote the development of SMEs, and all innovation for smart villages and that will be the *smart solution*. We can say that there is no smart villages without SMEs and there is no local solution without SMEs.

Village Angels

It is necessary to create 'village angels' and transform their 'business angel' role. This new initiative promises to boost more innovation through smart solutions, in which SMEs can be directly involved. Furthermore, this new initiative should play a crucial role in reshaping the activities of these enterprises. Regarding the complexity of the smart villages approach and its local focus, we can clearly conclude that it gives a level playing field for both local communities and local SMEs.

'Village angel' can be a company, an investor, or a person who create smart solution and develop smart villages. It is not just an application, but it should be more. They can create smart solution in the area of services, transport, local life, education, healthcare, etc. SMEs can create most of the innovation, and innovation can support quality of life.

Village angels will be the link between businesses, local communities and administrations. The EU's current rural development policies are concentrating merely on agriculture and agri-food activities and having less attention to social aspects. Therefore, the involvement of mayors is of utmost importance. They possess an enormous reservoir of clever solutions with no organized exchange networks. They accumulate trust; however, in most cases, this social value remains blocked within their respective settlements. In order to unblock the potential of their practical knowledge, a new type of cooperation mechanism is needed, which can be a well-managed database of local solutions, especially in the field of energy, water, mobility, housing, education, and resource-handling. If it becomes a real and practical solution, then that is the entry point for SMEs.

Sustainability and Hubs

The system that should be created around smart villages needs to be sustainable. It must provide a solution for all areas of life and has to be flexible. Easy and fast applicability and adaptability are strong expectations toward smart solutions. The standardization of quality features and rules is of critical importance, which facilitates the rapid implementation of the different solutions. We are mindful of the fact that there are no 'one size fits all' solutions; however, we have to make sure that applications developed for certain areas can be easily adopted in other regions or countries as well.

By doing so, standardization can provide a basis for linking smart villages within and even between countries. By connecting the system, a smart network can be created. Thus, the rural areas can be connected to each other and to big cities, too. A solution will become 'smart' if it takes advantage of the opportunities given by digital evolution, is Internet based, and most importantly can be used with smart tools (phone, tablet, and computer); that is, if it has the potential to help rural areas in their digital catching-up. An important criterion is that services must be available and accessible to everyone. EU funds must be used for this purpose.

In brief, the role of SMEs is based on three components in the light of intelligent or smart solutions:

(1) provision of services – development and customization of applications;
(2) service management – maintenance, improvement, adaptation, updating; and
(3) providing access to the services for all.

The applications would provide important information about the life of the village — regarding, for instance, pharmacy, doctor, police, cultural events etc., thus facilitating and modernizing the rural life. The aim is that smaller settlements could keep pace with development, too.

Local needs require real local solutions, while exogenous interventions should be limited. This notion is confirmed by the experiences of the earlier EU funds and other interventions, when companies tried to abuse the rules and loopholes and made investments and created jobs in cities using the resources devoted to rural areas. Let us have a look at some of the innovative developments demonstrating that small business and smart projects can deliver innovative, well-used solutions to local needs. The goal is that these solutions should be developed in a guided, organized way in accordance with the smart villages concept. For the system to work well, solutions must be easily copied and adaptable. To this end, an independent European tender and selection methodology should be set up, in order to avoid unnecessary overlaps.

Finance

At the EU level, there are several programs offering sources of financing for the SME sector. But there are no real synergies, with no real focus on the rural areas. The EU supports businesses and entrepreneurship with a number of measures. It also pays particular attention to SMEs, as their role in the European economic mechanism and job creation is outstanding. Exactly for this reason, the EU considers funding programs and applications as a policy priority to help companies get access to resources. In theory, there are many funding opportunities for SMEs: structural and investment funds, tenders, and loans, although still a lot has to be done in the field of microfinancing.

The EU's SMEs policy focuses on access to finance, market access, competitiveness, and innovation. Its aim is to promote entrepreneurship and create a business-friendly environment. The small business package of the EU should list the following principles.

It is widely understood that most SMEs are confronting serious limits in geographical, cultural, and thus economic terms. Economies of scale are obviously challenged in rural areas especially in the case of local services, like shops, post offices, schools, and so on. Improvement of mobility and smart provision of these services are equally welcome. But the question remains, namely who is paying for the difference. The answer is as follows: if not the market, then the public.

Prevailing EU policies are favoring robust interventions with the hope that massive one-time financial provisions may create self-sustainable solutions. Reality is different. Let's look at a simple example of air conditioning. Back in times, natural ventilation was the way to have fresh air in the buildings and constructing activities reckoned with the nature of airflows, including the heating systems. If designed and positioned well, the result was a complex and sustainable air-management, obviously with all its other archaic shortcomings. Today, external circumstances are having less importance and architects put in place electrical air conditioning installments maintaining the temperature level. Say smart. But if you cut electricity or other artificial energy supplies, the building becomes an unbearable environment, there is no hope for self-sustainability. The same applies for assistance programs. If we start an intervention, the players will accommodate themselves accordingly; therefore, in order to meet the projected goals, the financial tools have to be kept online for a longer run.

Financing smart villages appear to be a never-ending financial exercise; however, gains are also constant in the form of population balances and cultivated environment. The service that smart villages offer to the rest of the societies is their existence with a lot of values added, like educated and empowered people in the countryside who can generate additional incomes and revenues, like post offices in small settlements. The question is normally not about economic feasibility or profitability but the provision of access to same basic services for people living around, enabling them to catch up with cities, at least to a certain extent.

The various development programs of the EU are all having more or less attention to rural areas but in most cases with an auxiliary manner. In general, such elements are integrated into the specific programs, but do not meet each other locally. The financial part of the smart villages is supposed to be rather a managerial software giving interlinks or one-stop-shop opportunities especially to SMEs, like a local small company that intends to merge, for instance, limited amounts of educational assistance, energy-saving activities, and water supply. Today, it needs to apply for the relative funds one by one, whilst in a projected smart future we should pave the way for small but complex access.

Analysis Some Projects

Analyzing the webpage of The European Network for Rural Development (ENRD), they made a special 'Smart Villages Portal' (European Commission, 2018e). There is many inspiring projects and initiatives illustrating different aspects of the possibility of smart villages and, in particular, rural development. It's good to see and it can give confidence that there are some good projects which work to improve rural services such as healthcare, education, energy, and mobility. They try to build their own community as a smart village. This is the first step toward the goal. We should analyze these projects and draw the conclusions and, of course, to use the most relevant and important elements of this (European Commission, 2018f) In these projects, the positive and similar fact that they try to build own small microeconomy, they find some special local solution, their community is sustainable, and they want to go with the world.

The first project is a special small island, located off the West coast of Scotland. (European Commission, 2018g) The Isle of Eigg is one of the most beautiful Hebridean Islands. The island has a fascinating history, superb wildlife, and a vibrant community. The advantage of 'Eigg Heritage Trust' project is that they have found local achievements and solutions for most area of the life. The island is owned by the Isle of Eigg Heritage Trust who have managed it on behalf of the community since the community buyout of 1997. The Trust is the community organization which owns the Isle of Eigg. They have responsibility for stewardship of the island, its buildings and natural heritage, and for supporting future development. They manage their own agriculture; there are, for examples, three restructured farms; they pay attention to the environment; they provide the island with electricity from renewables. The idea is that it's not impossible to build fully renewably powered country. They think like a hub, and the main industries are tourism, agriculture, public services, construction, and the creative industries. They organize different activities to learn new skills, such as basket-making, felt-making, organic gardening, and so on. It's positive that they look for the partnership's possibilities and there is a volunteers' system with different opportunities. They have a strategy plan with their vision for the future development. They perfectly used the EU financing funds and programs. The result is that their population is sustainable, economical, developing, and happy (Eigg, 2018).

The second project is in Belgium. The 'Service hubs in rural Flanders' is one of the very interesting projects (European Commission, 2018h). They identify their challenges. These are very important steps that you can see clearly your state, difficulties, and challenges. They found that the solution may be a new concept, the 'Service Hubs'. For this, they identified three basic elements: (1) mobility; (2) services; (3) social; and also some additional elements what they needed: municipalities, local entrepreneurs, enterprises, organizations, local associations, and social innovation capitalists. Their main message is to reduce the necessity of functional trips to the city center and bring services back to the people. For the local solution they are perfect examples, and the major services are the local store, where the main concept that they should cooperate with local bakery, with local farmers, should be a focus for the regional products and performing small tasks for local inhabitants. Because smart village is not just an economic project but it should also be a social project. It focuses on the community center where everybody can find some activity. It can build sound communities, thus a good society. It finds good local solutions for the mobility hub, for example, with the system of cargo bike, car-sharing, and meeting points for functional trips.

Summarizing the experience, for a future smart villages, the main messages could be (1) to identify your challenges; (2) need a future vision; (3) think as a hub; (4) cooperate with inhabitants; (4) use renewable energy as far as possible; (5) be a social program; (6) co-working with local farmers; (7) find solutions for mobility; and (8) make local services. The projects' challenge is that they ought to use more smart devices because the main element of the new concept is to use the digital components and technologies.

Analyze Some Applications and Other Solutions

Let's mention a few successful ideas that can be interesting but we must emphasize that this is not really what we are talking about if we discuss about smart solutions for the smart villages. They just present that it is possible to develop smart solutions or solutions for rural areas. Our goal with these few examples is to draw attention to the fact that SMEs have the ability to develop smart solutions, and we just have to involve them and take advantage of their potential.

Mobility is one of the key problems in rural areas. There are some car share applications, such as 'Ring a Link' (Ringalink, 2018) This is a non-profit-making, charitable transport organization. They offer affordable and convenient transport primarily for rural dwellers. Their services allow travel to or from the local village or town for business, shopping, socializing, healthcare, or connection with national bus or train services. The service operates within areas surrounding large towns. These services have a schedule, but not a fixed route.

The Hungarian 'grandma application' is based on a social line. You can ask the older people – who are at home, sometimes alone – for some activities, if you need a cake or if you would like to learn sew and tie. The other Hungarian application 'our street' is based on a social line, too. On the app, you can see who you live around you and you can sell your products or your business. It may be a good solution in smart villages if you need some special services.

San-Car is a mobile dental healthcare service provided to rural communities in Romania and Bulgaria. Residents of isolated villages around the border between Romania and Bulgaria now have access to dental services often for the first time in their lives. A mobile care van from the SAN-CAR project provides free consultations and treatment to 14 isolated villages without dental medical services, improving inhabitants' health and quality of life. The project provided real social benefits for rural communities by improving access to medical services. The mobile dental clinic improved the health of people from isolated rural areas and increased their quality of life. (European Commission, 2016)

A Hungarian village has its own app. The inhabitants and visitors can get important information on special cultural events around. On the app you see details of how the village and local services go. On 'Ceglédbercel', there are many Wi-Fi hotspots, and a security camera system provides for peace (Egov, 2018).

We may read "The 25 Most Innovative AgTech Startups in 2018" (McGrath, 2018) too. They collected the most interesting innovative startups in the ag-tech space. This year's list features 12 alumni and 13 newcomers. Just, for example, using machine learning and data analytics, the company has compiled a database of which types of microbes work best to promote higher crop yields, or the other company is an indoor farming company utilizing machine learning, artificial intelligence and crop science to optimize yields and give produce exactly what it needs to achieve optimal freshness and taste (McGrath, 2018). It demonstrates that there are a lot of possibilities and there are a lot of capacities for smart innovation including agriculture, which is of top important for the future.

Conclusions and Recommendations

Deprivation and depopulation of rural areas is a piercing phenomenon, across the Globe. Despite the relative availability of modern technologies, rural settlements are becoming more and more isolated from the cultural, educational, and social point of view, while the available every-day services are getting even poorer. This process endangers us by turning into a spiral move. Moreover, natural devastation is another consequence due to harmful agroindustry practices and excessive use of chemicals, plastics etc. We must overcome this. We should find an alternative future for rural Europe. By its nature, rural life is not a matter of simple profitability considerations but a socio-ecological challenge. Small settlements must not become battlefields of giants; therefore, exclusively small industries and services may get a role. In other words, rural Europe should be a safe and protected haven for SMEs of any kind.

Digital revolution has different appearances in the various segments of our societies. Its impact results in good and bad; thus, humans are expected to regulate life accordingly. The digital divide is widening between small villages and big cities. Solidarity is a matter of clear interest. Villages are not to be abandoned, but urban gains have to be shared with those who provide us with the most important services by producing healthy food and guaranteeing shelters and rescues in the countryside. The intended spread of precision agriculture, research for modern alimentation, and rendering vast services in rural areas call for the promotion of smart solutions, startups or, in other words, SMEs. If urban achievements are cut off, they will die. If small settlements are not plugged in, they will die, too. Due to heavy investments, urban areas are kept online while rural Europe is struggling for having access to at least a minimum of public services. The smart villages' concept intends to award rural population with a chance to become able to keep pace with technological advancements and at the same time being capable of rendering all expected services like quality food, protected environment and offering a sound human living space. Apart from the necessary investments, other incentives might be considered as facilitated taxation schemes. However, sustainability does not prevail in itself. Tailor-made investments in knowledge or modern technologies should be continuous, whereas one-time financial injections are in vain. There must be a way in order to gather all good practices and implementing them in our rural Europe, which is still the background of half of the EU population.

Every smart village has to find its own uniqueness and take its potential and advantages. Saying more, it is simply not an economic project, but a political and a social one. It is little to say that there are many institutional European financing possibilities clearly dedicated sources are needed. Apart from seizing fresh resources existing ones are at help, too. Most of the available financial instruments are having certain rural and/or smart targeted provisions. The challenge is how to identify and ring-fence such elements, how to blend them and, finally, how to make such framework operational via a regulative 'software'.

Having regard all above it seems essential to acknowledge the very specific manner of rural areas in which most of the business, environmental and social services may not be rendered by others than small enterprises; thus, the European policy agenda has to realize its special responsibility to design particular frameworks for the sound cooperation of all players. To this end, there are some recommendations at the policy level: First, all villages are smart except there are a few a bit smarter. To become a smart village, first we need adequate local hope, ambition, and openness, with a strong commitment toward SMEs. Second, we need a perspective vision and rolling planning for at least 10 years what we want to achieve in technological and social terms, clarifying the role of small businesses. Third, during the next phase of the Commission's action-plan the identification, ring-fencing, blending and interlinking maneuvers must be a priority, with special focus to existing schemes targeting SMEs. Finally, political dedication has to be maintained both in merit of smart settlements including smart agriculture and smallholdings.

References

Egov. (2018). Smart villages, smart settlement, digital citizes. *E-Government, Hungary*. Retrieved from https://hirlevel.egov.hu/2018/06/09/okos-falvak-okos-telepulesek-digitalis-kisvarosok-a-lathataron/ Accessed on October 23, 2018.

Eigg. (2018). *About Eigg: The Isle of Eigg official portal*. Retrieved from http://www.isleofeigg.org. Accessed on October 23, 2018.

European Commission. (2015a). *Entrepreneurship and small and medium-sized enterprises (SMEs)*. Brussels: European Commission. Retrieved from http://ec.europa.eu/growth/smes/. Accessed on September 10, 2018.

European Commission. (2015b). *Annual Report on European SMEs 2013/2014: A Partial and Fragile Recovery*. Brussels: European Commission. Retrieved from http://bookshop.europa.eu/en/annual-report-on-european-smes-2013-2014-apartial-and-fragile-recovery-pbET0415079/. Accessed on September 15, 2018.

European Commission. (2016). *EU regional and urban development, regional policy, projects, SAN-CAR – Mobile dental healthcare provided to rural communities in Romania and Bulgaria*. Brussels: European Commission. Retrieved from http://ec.europa.eu/regional_policy/en/projects/bulgaria/san-car-mobile-dental-healthcare-provided-to-rural-communities-in-romania-bulgaria. Accessed on October 23, 2018.

European Commission. (2017). *EU Action for smart villages*. Brussels: European Commission. Retrieved from https://ec.europa.eu/agriculture/sites/agriculture/files/rural-development-2014-2020/looking-ahead/rur-dev-small-villages_en.pdf. Accessed on September 8, 2018.

European Commission. (2018a). *European network for rural development, smart villages portal. Brussels: European Commission*. Retrieved from https://enrd.ec.europa.eu/smart-and-competitive-rural-areas/smart-villages/smart-villages-portal_en. Accessed on September 8, 2018.

European Commission. (2018b). Internal market, industry, entrepreneurship and SMEs. *What is an SME?* Retrieved from http://ec.europa.eu/growth/smes/business-friendly-environment/sme-definition_en. Accessed on September 10, 2018.

European Commission. (2018c). *Horizon 2020 Work Programme 2018-2020: Innovation in small and medium-sized enterprises.* Brussels: European Commission. Retrieved from http://ec.europa.eu/research/participants/data/ref/h2020/wp/2018-2020/main/h2020-wp1820-sme_en.pdf. Accessed on September 16, 2018.

European Commission. (2018d). *Policies, information and services: Horizon 2020: Innovation is SMEs.* Brussels, European Commission. Retrieved from https://ec.europa.eu/programmes/horizon2020/en/h2020-section/innovation-smes. Accessed on September 18, 2018.

European Commission. (2018e). *European network for rural development: Smart villages portal.* Brussels: European Commission. Retrieved from https://enrd.ec.europa.eu/smart-and-competitive-rural-areas/smart-villages/smart-villages-portal_en. Accessed on October 22, 2018.

European Commission. (2018f). *European network for rural development: Smart villages portal, projects and initiatives.* Brussels, European Commission. Retrieved from https://enrd.ec.europa.eu/smart-and-competitive-rural-areas/smart-villages/smart-villages-portal/projects-initiatives_en. Accessed on October 23, 2018.

European Commission. (2018g). *European network for rural development: Smart villages portal, projects and initiatives, Eigg Heritage Trust.* Retrieved from https://enrd.ec.europa.eu/sites/enrd/files/s4_rural-businesses_eigg_bryan.pdf. Accessed on October 23, 2018.

European Commission. (2018h). *European network for rural development: Smart villages portal, projects and initiatives, service hubs in rural Flanders.* Retrieved from https://enrd.ec.europa.eu/sites/enrd/files/tg2_smart-villages_service-hubs_hoet.pdf. Accessed on October 23, 2018.

Eurostat. (2013a). *Glossary: Innovation-active firm.* Retrieved from https://ec.europa.eu/eurostat/statistics-explained/index.php?title=Glossary:Innovation-active_firm. Accessed on September 20, 2018.

Eurostat. (2013b). *Glossary: Innovation activity.* Retrieved from https://ec.europa.eu/eurostat/statistics-explained/index.php?title=Glossary:Innovation_activity. Accessed on September 20, 2018.

Eurostat. (2017). *Innovation statistics.* Retrieved from https://ec.europa.eu/eurostat/statistics-explained/index.php/Innovation_statistics. Accessed on September 20, 2018.

Lytras, M. D., & Visvizi, A. (2018). Who uses smart city services and what to make of it: Toward interdisciplinary smart cities research. *Sustainability, 2018*(10), 1998, doi:10.3390/su10061998

McGrath, M. (2018). The 25 most innovative AgTech startups in 2018, Forbes, 27.06.2018. Retrieved from https://www.forbes.com/sites/maggiemcgrath/2018/06/27/the-25-most-innovative-agtech-startups-in-2018/#420226e22302. Accessed on October 23, 2018.

Ringalink. (2018). *About ring a link. Ring a link offical portal.* Retrieved from http://www.ringalink.ie. Accessed on October 23, 2018.

Szanyi-Gyenes, X., Mudri, Gy., & Bakosné, B. M. (2015). New tools and opportunities in growth and climate friendly greening for small and medium enterprises in the European Union, Budapest: APSTRACT – Applied Studies in Agribusiness and Commerce (Vol. 11, Number 4). Retrieved from https://ideas.repec.org/a/ags/apstra/226118.html. Accessed on September 09, 2018.

Visvizi, A., & Lytras, M. D. (2018a). Rescaling and refocusing smart cities research: From mega cities to smart villages. *Journal of Science and Technology Policy Management (JSTPM)*, 9(2), 134–145. doi:10.1108/JSTPM-02-2018-0020

Visvizi, A., & Lytras, M. D. (2018b). It's not a fad: Smart cities and smart villages research in European and global context. *Sustainability*, 2018, *10*(8), 2727, doi:10.3390/su10082727

Chapter 9

Smart Villages in Slovenia: Examples of Good Pilot Practices

Veronika Zavratnik, Andrej Kos and Emilija Stojmenova Duh

Introduction

In order to contextualize smart rural development in Slovenia, the main aim of this discussion is to present the current state of the art, address new interdisciplinary approaches in understanding the problems of rural areas, and most importantly, apply new approaches to smart rural development and introduce a new concept to the field. The objective of the chapter is multilayered. It starts by addressing the main problems of Slovenian countryside and its inhabitants. The review of initiatives and practices has already been made elsewhere (Zavratnik, Kos, & Stojmenova Duh, 2018), and we will only summarize it briefly to emphasize the most important achievements/improvements. The chapter further focuses on the state of the art in Slovenia in connection with smart and sustainable rural development, and analyzes the results contextually. Three promising solutions/practices that local communities found to answer the existing challenges in different sectors (i.e., tourism, mobility, and innovation) are presented. Most findings are based on the fieldwork, visits to the sites, and work with the local communities. In the last part of the chapter we contextualize our findings and further explain our specific approach to smart rural development. Finally, the chapter introduces a new concept — that is, Smart Fab Village, which is built on the concept designed for urban areas and then carefully adapted for the needs and specific requirements of rural areas.

Problems of (Slovenian) Rural Areas

There are many somewhat romantic connotations to living in the countryside. These include: the sense of freedom that can be enjoyed in the midst of untamed nature and beautifully cultivated fields, breathing fresh air, growing your own food, getting freshly milked milk, or letting your children safely play in the front yard. The countryside also offers great possibilities for outdoor and sports activities, which has a positive impact on the quality of life of the people living there. But seen from another perspective, living in rural areas can also pose some challenges only rarely experienced in urban settlements, like limited access to welfare services, such as health or education. Living in rural areas can be especially challenging for the elderly and young people. For the former, the most challenging is the lack of healthcare and social services, as well as difficulties with transportation. Young people are having problems finding high-quality and long-term jobs outside of the agricultural sector.

These are not only regional or local problems, as the processes of urbanization are on the rise on a global scale and more and more people from rural areas are moving to live and work in cities and towns. A little less than half of the Earth's population now lives in rural areas and the numbers are in decrease (UN, 2014). Reasons for this situation are different and dependent on specific contexts, but broadly speaking the rural–urban migration is often the consequence of factors mentioned earlier. To be more specific, in Slovenia a little more than half of the population still lives in predominantly rural regions, and therefore it is very important to properly address the current state of the art of rural areas, to foster initiatives, and to develop new effective practices for smart rural transformation. Besides challenges connected with agriculture (such as land abandonment and enhancing agricultural productivity), one of the main components to be addressed within the national rural development plan is the creation of business opportunities other than those related to agriculture (European Commission, 2018a; MKGP, 2015).

Similar to other European Union (EU) regions, predominantly rural regions in Slovenia are shrinking. The phenomenon is best described with the combination of different factors: depopulation, that is, outmigration of young and educated people; aging of the rural population; emigration of well-educated individuals; and low diversity of job opportunities (ESPON, 2017, pp. 2–3). But there are some problems that are more specific for the Slovenian context. For example, given the specific relief factors, historical circumstances that led to fragmentation of land ownership and protection of areas with specific flora and fauna regulatory frameworks, for example, Natura 2000, the majority of family farms are small farms, which limits their prospects of growth and development and makes it harder to make a living only by agricultural activities. Further, in the light of the vision of the EU to give all EU citizens access to broadband internet, Slovenia is still a little below the European average concerning broadband coverage. According to the Digital Economy and Society Index (DESI), from 2018, 98% of Slovenian households have access to fixed broadband connections, but only 83% have access to fast broadband connections. This places Slovenia only to the 15th place among the 28 EU members (European Commission, 2018b). More specifically, concerning fixed broadband availability in rural areas, Slovenia is a little below the EU average (92.6 %) as well, reaching 92.2%. In this regard, the divide between urban and rural areas is evident, with rural areas still being below the average (European Commission, 2017a).

Because of the low population density, that is, on average 102.1 inhabitants/ km^2 (SURS, 2018) and dispersed settlements, rural areas are among the least covered in terms of broadband accessibility. Broadband accessibility has many positive economic and social impacts: it has a positive impact on employment rates and productivity, enables the provision of digital services, such as e-education and e-health, and has many positive environmental effects in terms of lower traffic and reduced carbon footprint (achieved by video conferences, working from home, etc.; MJU, 2016, pp. 4–6). Therefore, accessibility to high-speed broadband internet is the foundation of smart rural development. Conversely, the lack of it can have negative effects on providing better opportunities for rural citizens. And last but not least, Slovenian countryside is also suffering from poor traffic infrastructure (MKGP, 2018, p. 117).

All the aforementioned factors have a great influence on the development of rural areas; these are often perceived as places for vacations, resting, or recreation but only rarely as an environment for thriving SMEs, start-ups, or other innovative and creative business which produce high-added value.

Wider Framework and Its Implications for Slovenian Rural Communities

Tackling the challenges of rural areas has become one of the most important aspects of contemporary developmental policies and initiatives on a global scale. For example, in 2015, the General Assembly of the United Nations proposed 17 Sustainable Developmental Goals that are included in the 2030 Agenda for Sustainable Development (UN, 2018). Within this framework, the EU and its individual countries have also proposed and accepted various measures to make rural areas a more attractive environment to live and work in. More specifically, Slovenia has also played an active part in establishing and promoting the concept of smart villages within the national framework. But to understand the dynamics of the research field more properly, let us take a look at the wider framework of smart rural development.

Discussions on what a smart community is, how to understand the concept of smartness in view of developmental policies (Naldi, Nilsson, Westlund, & Wixe, 2015), and wider problems of research on smart villages have been proposed elsewhere (Visvizi & Lytras, 2018a), and therefore we will only expose some key points. First, there is not one single definition of the concept – the approach to smart development is always context-dependent and therefore conditioned by the problems of a local or regional community. In this context, it is necessary to apply bottom-up approaches. Second, although digitalization is a horizontal prerequisite for smart development, it is not the only one. Social and cultural contexts play a very important role in the understanding and successful application of new approaches and initiatives in practice. Visvizi and Lytras (2018a, p. 2) argue that because research on smart villages is only at the beginning, there is a great opportunity to avoid the so-called "ICT-hype" and focus on the problems that villages face in the first place. Third, smart development approaches always put sustainability at the forefront of the community (Zavratnik et al., 2018).

Since 2016, when the Cork 2.0 Declaration was adopted by the EU European Commission (2016), smart villages have become present in many developmental debates in Europe. But speaking globally, the concept is actually not altogether a novelty. There are many global-wide initiatives concerning this matter, for example, Smart Village Initiative (Smart Village, 2018) or Smart Village priority program of Institute of Electrical and Electronics Engineers foundation (IEEE Smart Village) (IEEE Smart Village, 2018). Many times, accessibility to electricity for off-grid communities is at the forefront of activities in these organizations. These initiatives are very important in the context of tackling the challenges of global inequality, but the Slovenian context is much more dependent on specific initiatives and regulations of the EU.

Cork Declaration (European Commission, 2016a) paved the way for discussions on smart rural development in 2016. In 2017, the European Commission initiated EU Action for Smart Villages (European Commission, 2017b), where it addressed many fields, such as agriculture, mobility, social innovation, and the digital divide between urban and rural areas, which also reflected on the villages of the future. In April 2018, the EU launched Bled Declaration, under the name Smarter Future of the Rural Areas in EU (Bled Declaration, 2018). The latter is especially important for Slovenia as it was signed in Bled, Slovenia, which puts Slovenia not only implicitly but also explicitly on the map of smart rural development initiatives.

In addition to declarations and actions to accelerate rural development, the EU has also proposed other important policies. Rural Development Policy 2014–2020 proposed various national development programs and enabled the implementation of the LEADER program, which has recently been extended to Community-Led Local Development (CLLD). In its essence, this is a bottom-up approach where the main focus is being put on local communities and promotion of new solutions. In LEADER and CLLD approaches, an important role is also played by Local Action Groups (LAGs) that enable cooperation on different (inter)national levels. ESPON 2020 Cooperation Programme is yet another program aiming at reinforcing the cohesion policies and promotion of the territorial dimension in developmental practices. Within ESPON, the main target groups are policymakers and national/regional/local authorities. Its main objective is to enhance applied research, upgrade the transfer of knowledge, improve territorial observation, and make the implementation of provisions more effective (European Union, 2016). Within these and some other policies, various projects on smart development have been proposed and implemented (Zavratnik et al., 2018, pp. 5–8).

For the Slovenian context, it is very important that the first activities to promote the concept of smart village and inform the public on the importance of smart and sustainable development were organized already in 2017. The main ambassador of those events in Slovenia and also at the EU level is Franc Bogovič, a Slovenian member of the European Parliament. By now, 18 round tables have been organized in different parts of Slovenia, almost always in very small, rural communities. Participants were very diverse: from IT experts, developmental agencies, farmers, mayors, to electrical engineers (Bogovič, 2018). Therefore, local communities have been actively involved in the discussions on smart rural development and have had the opportunity to hear how the smart village concept would be applied in their local context.

Examples of Existing Practices

First, there are no better words to describe Slovenia than the phrase 'Green Country': almost 60% of Slovenia's landscape is forests, and this puts it to the third place in the EU by the percentage of woodlands (ZGS, 2018). Second, according to urban–rural typologies (NUTS 3), all of the 12 Slovenian regions are either predominantly (e.g., nine) or moderately (e.g. three) rural (SURS, 2017). There are actually no urban regions in Slovenia (Eurostat Statisctics Explained, 2013). Therefore, as already stated, forests represent around 56% of the territory, whereas agricultural territory accounts for almost 40% (European Commission, 2018a). Third, although being a very urban settlement, Ljubljana, the capital of Slovenia, was pronounced the green capital of Europe in 2016 (MOL, 2018). With around 290,000 inhabitants, Ljubljana is the largest city in Slovenia, but has nevertheless not lost its 'green' nature or its sustainable orientation for the future. For example, in 2014, Ljubljana joined the network Zero Waste Europe and became the first European capital to join in (MOL, 2014). Given the importance of green and rural areas for Slovenia, it is necessary to put the concept of smart villages into practice. To do so, it is crucial to start building on existing initiatives and practices.

As the new paradigm of rural development has shifted attention from the agricultural sector to the broader concept of rural space (Lorber, 2013), our attention is also very broad. In the context of smart rural development, existing local/regional practices and initiatives on rural development are documented and monitored through the process of working together with the communities. We also aim to evaluate these practices in the light of sustainable and smart development. The examples we are presenting here and also the whole areas included in our broader research were chosen on the basis of their specific local/regional problems and existing initiatives; low tourist development, mobility problems, aging of the population, and scarce innovation opportunities.

Tourism

First, it has already been noted that rural tourism is one of the key drivers of rural development (Hernández-Maestro & Gonzalez-Benito, 2013) also in Slovenia, and therefore we have examined the area more thoroughly. One of the practices that has already been implemented is distributed hotels. A distributed hotel is an innovative form of cooperation in the field of tourist accommodation. Usually it connects different owners or managers of accommodation facilities that manage the hotel together. The hotel consists of one accommodation entity, but apartments and rooms are located in various buildings across the village/town but with one central place – reception, restaurant, and the like (Confalonieri, 2011). In this way, smaller communities that otherwise do not have the capacities to develop tourist activities on a large scale can develop those activities on a smaller and more locally embedded scale. Many times, distributed hotels include various segments of a community: local farmers, craftsmen, providers of other tourist activities, restaurants, food producers, etc. (Vrbica, Šifkovič, 2017).

For example, the first distributed hotel in Slovenia, Distributed Hotel Konjice, opened its doors in January 2018 and is closely connected to its local community; by now it connects three providers of tourist accommodation – Mansion Trebnik, Zlati Grič (Eng. Golden Hill), and Mala kmetija (Eng. Small Farm). The municipality is the owner and the manager of the main building, Mansion Trebnik, and a strong supporter and promoter of the hotel. In the restaurant, which offers breakfast and lunch, only locally produced food supplied by local providers (bakers, cheesemakers, farmers, etc.) is served. There are strong connections with the local high school center, which places its international guests in the distributed hotel. An interesting connection between the school center and the hotel was made when school's FabLab (i.e., U-Lab) produced number plates for the rooms. Providers of other tourist services also collaborate with the hotel; the strongest connections are established with the ski resort Rogla and Carthusian monastery Žiče in the vicinity. In Mansion Trebnik, there is also a conference room that can be rented, and a café Mali grof (Eng. Little Count). Although the distributed hotel in Slovenske Konjice was opened very recently, the number of guests per month is steadily increasing and reached 50 in August 2018. Distributed Hotel Konjice is the only provider of tourist accommodation in the local area (Razpršeni hotel Konjice, 2018).

Another distributed-hotel-in-the-making is located in the Slovenian coastal region, in the village Padna. Padna is a small settlement, counting only around 180 inhabitants but with an already well-developed tourist activity (Istrske hiše Padna, 2017). It is located in the backcountry of the Slovenian coast, Slovenian Istria. In 2012, the village received the first tourists from cruise ships that landed in Koper harbor nearby. Since then, tourist activity has been actively developed by locals; old buildings have been renovated to increase the accommodation facilities and host the local tourist information center. Parallel to the development of tourism runs the promotion and restoration of local cultural heritage, such as the renovation of an old water fountain and paths leading to it. As tourist activity has been present here for some years and is now very well established, local initiative for tourist development is very strong. From around 150 tourist visits before 2012, now there are 6,000–8,000 people visiting Padna yearly and more than 1,500 people stay there overnight. A more steady and higher number of overnight stays would bring opportunities to develop new services (like a local market, store, or new mobility opportunities), but for the future development of accommodation capacities it is important to note that the Istrian village Padna is part of protected cultural heritage and thus subject to strict rules for the protection of cultural heritage. The only way to offer new accommodation capacities for tourists is to renovate existing buildings and rearrange them into tourist accommodation sites. For Padna, strict rules apply for renovation and building. The project Distributed Hotel Padna has already been proposed, but its realization depends on finding an investor that would be willing to enter a private–public partnership with the Municipality Piran. The project proposed 10 new accommodation sites to be offered. Altogether this would provide capacities for around 150 people. In general, tourist activities already connect different sets of local actors and therefore the idea of a distributed hotel is very well accepted among the locals. Through developing sustainable practices that enhance the growth of local economy, the future vision of Padna is a smart village.

Because of the vicinity of the sea and willingness of the inhabitants to participate, Padna has a great potential to further develop its tourist activities. The most western Slovenian village Robidišče, in contrast, has only seven permanent residents. Even a few years ago, most houses in the village were destroyed and unsuitable to live in, but today there are some initiatives on tourist development. For example, there is one provider of accommodation and one organic farm, and even though they are not formally connected in any way, they work and coordinate some activities together. In this year, a third actor started the renovation of a house to establish a small hotel to be opened in 2019. The three of them have also established a relationship with the municipality. Robidišče has an initiative on the establishment of a distributed hotel, but it has not yet been realized. Even though the initiative has not been formally realized yet, on the conceptual level, the partners already use the concept of a distributed hotel; they help each other with the organization of events, borrow each other the capacities, use local resources, and develop the concept of sustainable tourism.

In the Slovenian region of Karst, another example of accommodation opportunities is based on the model of a distributed hotel – Hotel of Good Teran (Agrotur2, 2018). The initiative was launched by the Agricultural Institute of Slovenia within the Interreg Italia-Slovenia projects Agrotur I and Agrotur II, and connects those producers of the wine Teran from both sides of the border that also offer accommodation possibilities. It is managed by the Association Consortium of Karst Teran Makers; there are 17 providers of accommodation included in the initiative. At the moment, the hotel is working only on an online platform, but there are only a few actual connections between the providers: even though the providers are formally connected into an initiative, most of them are working on their own, using their own promotion channels, with their only form of collaboration being the online platform. Therefore, at the practical level, the hotel actually does not exist as there is no mutual collaboration or consensus on what it means to be included in the hotel, and the role of the manager is purely formal, not implemented in practice.

There are some other local initiatives on the tourist development of rural areas – for example in Blatna Brezovica, a village in the proximity of Ljubljana, or in the Municipality of Žalec – but they are in a very early development stage.

Mobility

The next very intriguing practice is connected with mobility. In Slovenia, mobility in rural areas is explicitly connected with personal/family cars. For the elderly unable to own or drive their own cars, it is therefore very challenging to have access to some basic services – health institutions, administrative units, stores, a market, etc. Therefore, there is a strong need to resolve problems of great car dependency in rural areas. For the future, one of the promising projects seeking to offer sustainable solutions is a project Sustainable Mobility Behaviours in the Alpine Regions (SaMBA), funded by Interreg Alpine Space, trying to find alternatives to personal cars and offering solutions based on public transportation (LTFE, 2018). In terms of existing practices, the Institute Sopotniki (Eng. Co-travellers) was established in 2014, providing the elderly with the opportunity to activate their social life and enhance intergenerational solidarity. Free transport services enable access to basic services and social activities the elderly want to visit. It works on a voluntary basis and includes 23 volunteers and 110 regular users (IPOP, 2016). We have detected a very similar practice in Kungota, a municipality in northeast Slovenia with a little fewer than 5,000 inhabitants. The municipality has bought a car that is used by "young pensioners" who offer voluntary mobility services to those who don't have or are unable to drive a car. The users of the services pay for the gas, and the municipality takes care of the car. In Kungota, this kind of intergenerational cooperation is further enhanced by activities organized by the House of All Generations. The house hosts cooking classes, board game nights (organized by youngsters), bowling, yoga lessons, and many others. The participants also cultivate a community garden, and use vegetables to cook joint lunches once a week and make their own herbal tea.

Innovation

The next example of a promising initiative existing also in rural areas is FabLab Network Slovenia. The Network is a good option to interconnect in some way all the aforementioned initiatives. It connects more than 80 partners from different sectors, among them 28 FabLabs (Mreža FabLab Slovenija, 2018). The name FabLab is an abbreviation for a *fabrication laboratory*. Therefore, FabLab is an open, creative coworking place where everyone can learn, invent, play, or develop their (projects) with the use of digital technologies and with the help of mentors. FabLabs are equipped with the latest technological equipment connected to their local environments, and were built on the basis of local initiatives. Some FabLabs in the network are located in smaller communities in the rural environment and are therefore even more strongly intertwined with the local community. Such smart, innovative places represent an opportunity for rural youth to gain essential digital skills important for the future labor market. Nevertheless, their potential is only rarely used by the local community. For example, FabLabs could be places where students and others could gain qualifications and skills needed by local companies; these are places that stimulate innovation and creativity, which also improves opportunities for new business ideas, entrepreneurship, and intergenerational collaboration in terms of exchange of skills and knowledge (Stojmenova Duh & Kos, 2016). FabLabs represent a coworking place for young start-ups or freelancers to collaborate and work together, and can therefore become a platform in enabling higher diversity of job opportunities. Further, offering new job opportunities, FabLabs could also be places for enhancing rural–urban linkages in terms of other than agricultural goods, and therefore contribute to reducing the depopulation of rural areas.

Proposing New Concepts: What Is a Fab Village?

The practices and examples described earlier offer a starting point for a discussion on the future of smart rural development in Slovenia; building on existing practices provides the best options for new initiatives to succeed. In an ongoing research on rural development, existing examples of successful practices are being identified and combined with our expertise in the field of digitalization and information and communications technology (ICT). In this regard, we propose a new developmental concept to be applied in rural areas: Smart Fab Village. Smart Fab Village is an innovative combination of principles and practices on three different levels: local initiatives, Smart Villages, and Fab Cities. In this view, smart rural development is conceptualized on the idea of a smart village, while using local initiatives to enhance circular economy, use and development of local (natural, cultural, and social) resources, and economy. A similar view is also the foundation for understanding the concept of Fab Cities: they have developed as a concept corresponding to urban challenges, as an answer to environmentally exhausting ways of contemporary cities. Fab Cities concentrate on the production and consumption of things, while the main aim is to redefine relationships between how things are made and advanced digital technologies. Fab Cities are built upon the ideal of FabLabs: connectivity, creativity, and culture. The vision for the future is that cities would produce almost everything citizens need inside its borders – including recycling waste material, and innovatively transform it for new uses and purposes. In the center of this vision are FabLabs as enablers of small-scale, locally based, and sourced production that is environmentally friendly and sustainable in its approaches. For this reason, the Fab City Global Initiative and the global FabLab Network are inextricably linked (Fab City, 2018).

As predominantly rural, Slovenia has no initiatives on the Fab City movement. The Slovenian national FabLab Network, in contrast, covers a larger part of rural areas and therefore offers a springboard for thinking about how to apply the Fab City principles in rural regions and interconnect them with existing practices/initiatives to make them even more effective and sustainable. Therefore, the newly developed concept Smart Fab Villages is proposed as an example of how the ideals of the concept developed in and for urban regions can be applied in rural regions if using appropriate infrastructure. We see Smart Fab Villages as an approach to rural development that uses the network of FabLabs to build upon and enhance the existing practices on rural development in order to build up the sense of community, local and circular economy, foster innovation and creativity, diversify job opportunities, and strengthen sustainable orientation of rural development practices.

In our view, FabLabs are the main interconnecting points between the processes of digitalization and solving of problems that communities face. A promising potential of digital technology and ICT in urban areas has already been discussed elsewhere (Ringenson, Höjer, Kramers, & Viggedal, 2018), and the main aim of this chapter is to emphasize its importance also for rural areas. FabLabs are incubators for gaining and developing (digital) competences, and in rural areas, they are one of the prerequisites for opening the way to the empowerment of rural inhabitants, especially in the fields of technology and digital innovation.

Some of the founding principles of FabLabs are collaboration, open innovation, and mentorship, and the main idea of FabLabs in rural areas is to strengthen the economic and social conditions of the communities (Stojmenova Duh & Kos, 2016, p. 214). By enhancing the principles of open innovation, there is a greater possibility also for social innovation. And, by increasing digital competencies, the opportunities for developing new (digital) services – connected with existing practices – emerge. From this perspective, there are great possibilities on how to further develop practices on mobility or tourist development by using ICT. For the future of distributed hotel in Padna, the vision is to make a smart hotel based on an application that will enable check-in and check-out without visiting the main hub of the hotel but simply by using the application on site. To illustrate further on the example of Slovenske Konjice: the existing connections between the distributed hotel and the FabLab are a great starting point. Further, with FabLab as a focal point, new digital solutions and strategies, web applications, or websites/stores could be developed to facilitate the use of services for tourists and make coordination easier and more effective. FabLab could also be the place where local craftsmen like carpenters or jewelry designers would manufacture their products.

Conclusions and Recommendations

As an interdisciplinary group of researchers from different fields of social and natural sciences, we aim to gain a more holistic understanding of the problems and challenges of rural areas. The field research carried out in the Slovenian rural context is one of the steps in this direction. Based on the experiences, this chapter has shown how pan-European problems are made explicit within the Slovenian context and how local communities have developed their very own contextually conditioned solutions. Slow tourist development is being encouraged by the establishment of distributed hotels. The problems of mobility of older generations are being addressed through practices based on voluntary mobility services. The lack of opportunities for the young and consequential 'youth drain' of Slovenian rural areas are addressed through the establishment of FabLabs embedded in local rural environments. Aging of the population is addressed within intergenerational centers and cooperation of different stakeholders. Most importantly, many community-led initiatives are sustainable oriented, considering local/regional economic, social, and cultural contexts.

Combining multidisciplinary knowledge and academic expertise in the field of digitalization/digital transformation of rural areas and ICT, the existing initiatives and practices were used to develop new approaches to smart rural development. These approaches are built on cocreation and cooperation with communities. Similar points on inhabitants/citizens involvement and also on the role of ICT have already been made for urban contexts by Visvizi and Lytras, who have also exposed the importance of existing correlations between urban and rural areas (2018b). In the context of Slovenia, FabLabs are seen as focal points for further sustainable development and diversification of rural economy. These places of innovation and creativity could and should be the meeting points connecting different generations, incubators for new business ideas, accelerators of digital literacy, and places of sustainable small-scale and locally contextualized production. In this way, social, cultural and natural local resources would be used more sustainably and included in the processes of circular economy. Another important objective of seeing FabLab as an important part of smart rural development is intergenerational cooperation and mentorship that would accelerate the transfer of different important knowledge contributing also to (social and digital) innovation, economic development, and stronger community building. In this way, a local FabLab can be intertwined with every aspect of the society in order to enhance local economy and societal development.

The concept of Smart Fab Villages proposed here uses the existing infrastructure and acknowledges local and regional needs by using a bottom-up principle, and sees digitalization as a horizontal prerequisite to social and economic development.

Considering the processes of rural shrinkage and depopulation of rural areas, it is now a critical time to change perceptions of rural areas as prospective and diverse places to live in. On the basis of an ongoing research on smart practices of rural development, the need for inclusive approaches that take into account the needs of rural communities is emphasized. In this context, more EU policy support is proposed for community-based bottom-up approaches that are led by the citizens. In addition, it is necessary to further develop existing linkages between urban and rural areas as they are inextricably connected. One of our main claims is that it is also necessary to increase funding for digital, creativity, and innovation skills development.

Only contextually developed practices based on real-life rural problems will be successfully integrated in everyday lives of rural citizens, and will therefore contribute to enabling equal opportunities for everyone, regardless of where they live – in a city or in the countryside. Therefore, the need to support the emergence of bottom-up and citizen-led approaches to contemporary issues is becoming of great importance. Stemming from this is also the necessity to include these citizen- and community-led initiatives into mainstream policy development and implementation processes.

Stemming from the proposal of the new Smart Fab Village concept developed from the concept adapted to urban spaces, there is an evident need for two future measures: measures for the promotion and further development of urban−rural and rural−urban linkages, and measures for the promotion of (digital, creativity, and innovation) skills development.

Acknowledgments

This chapter has been produced within the Smart digital transformation of villages in the Alpine space (Smart Villages) project, which was co-funded by Interreg Alpine Space (2018−2021). The authors acknowledge the financial support from the Slovenian Research Agency.

References

Agrotur2. (2018). Hotel of Good Teran: A boutique hotel in the Karst for special guests. Retrieved from http://www.hoteldobregaterana.si/en/About-us. Accessed on August 20, 2018.

Bled Declaration. (2018). *Smart villages: Bled Declaration*. Retrieved from http://pametne-vasi.info/wp-content/uploads/2018/04/Bled-declaration-for-a-Smarter-Future-of-the-Rural-Areas-in-EU.pdf. Accessed on June 2018.

Bogovič, F. (2018). *My work*. Retrieved from http://bogovic.eu/my-work/. Accessed on December 20, 2018.

Confalonieri, M. (2011). A typical Italian phenomenon: The "albergo diffuso". *Tourism Management, 32*, 685−687. doi:10.1016.

ESPON. (2017). *Policy brief: Shrinking rural regions in Europe − Towards smart and innovative approaches to regional development challenges in depopulating rural regions.* Luxembourg: ESPON EGTC.

European Commission. (2016). *Cork 2.0 Declaration: A better life in rural areas.* Retrieved from https://enrd.ec.europa.eu/sites/enrd/files/cork-declaration_en.pdf

European Commission. (2017a). *Broadband coverage in Europe 2016. Mapping progress towards the coverage objectives of the digital agenda.* European Union.

European Commission. (2017b). *EU action for smart villages.* Retrieved from https://ec.europa.eu/agriculture/sites/agriculture/files/rural-development-2014-2020/looking-ahead/rur-dev-small-villages_en.pdf

European Commission. (2018a). *Factsheet on 2014-2020 rural development programme for Slovenia.* Retrieved from https://ec.europa.eu/agriculture/sites/agriculture/files/rural-development-2014-2020/country-files/si/factsheet_en.pdf. Accessed on June 12, 2018.

European Commission. (2018b). *Indeks digitalnega gospodarstva in družbe 2018, poročilo održavi - Slovenija.* Retrieved from http://ec.europa.eu/information_society/newsroom/image/document/2018-20/si-desi2018-country-profile-lang_4AA75D84-E1F3-17B1-B011CC1513D6E43D_52354.pdf. Accessed on October 8, 2018.

European Union. (2016). *ESPON 2020 Cooperation Programme.* Retrieved from https://www.espon.eu/programme/espon/espon-2020-cooperation-programme. Accessed on October 12, 2018.

Eurostat Statistics Explained. (2013, February). *Archive: Rural development statistics by urban-rural typology.* Retrieved from https://ec.europa.eu/eurostat/statistics-explained/index.php/Archive:Rural_development_statistics_by_urban-rural_typology. Accessed on September 1, 2018.

Fab City. (2018). *Fab city whitepaper: Locally productive, globally connected self-sufficient cities.* Retrieved from https://fab.city/documents/whitepaper.pdf. Accessed on August 15, 2018.
Hernández-Maestro, R. M., & Gonzalez-Benito, Ó. (2013). Rural lodging establishments as drivers od rural development, *53*(1), 83–95.
IEEE Smart Village. (2018). *IEEE Smart Village.* Retrieved from http://ieee-smartvillage.org/
IPOP. (2016). *Trajnostna mobilnost v praksi: Zbornik dobrih praks.* Retrieved from http://ipop.si/wp/wp-content/uploads/2016/10/Trajnostna-mobilnost-v-praksi.pdf. Accessed on November 27, 2018.
Istrske hiše Padna. (2017). *Istrske hiše Padna.* Retrieved from http://www.istrskehisepadna.si/index.php?stran=padna&sklop=main&lang=si. Accessed on August 20, 2018.
Lorber, L. (2013). Spremembe v pristopih k razvoju podeželja - nova razvojna paradigma. *Revija za geografijo, 8*(1), 9–22. Retrieved from https://www.dlib.si/details/URN:NBN:SI:DOC-OL9TXS2J/?query=%27contributor%3dLorber%2c+Lu%C4%8Dka%27&pageSize=25
LTFE. (2018). *SaMBA – Sustainable mobility behaviours in the Alpine region.* Retrieved from http://ltfe.org/en/portfolio/samba/
MJU. (2016). *Načrt razvoja širokopasovnih omrežij naslednje generacije do leta 2020: Dostop do interneta visokih hitrosti za vse.* Retrieved from http://www.mju.gov.si/fileadmin/mju.gov.si/pageuploads/DID/Informacijska_druzba/NGN_2020.pdf. Accessed on August 15, 2018.
MKGP. (2015). *Program razvoja podeželja Republike Slovenije 2014-2020.* Retrieved from https://www.program-podezelja.si/sl/knjiznica/31-program-razvoja-podezelja-rs-2014-2020-osnovne-informacije-o-ukrepih-brosura/file. Accessed on August 5, 2018.
MKGP. (2018). *Program razvoja podeželja RS za obdobje 2014-2020.* Retrieved from https://www.program-podezelja.si/images/SPLETNA_STRAN_PRP_NOVA/1_PRP_2014-2020/1_1_Kaj_je_program_razvoja_pode%C5%BEelja/4._sprememba_PRP/PRP_Program_razvoja_podezelja_4_sprememba_2018.pdf. Accessed on August 15, 2018.
MOL. (2014). *Ljubljana v Zero Waste mreži.* Retrieved from https://www.ljubljana.si/sl/aktualno/ljubljana-v-zero-waste-mrezi/. Accessed on September 10, 2018.
MOL. (2018). *Zelena prestolnica Evrope 2016.* Retrieved from https://www.ljubljana.si/sl/moja-ljubljana/zelena-prestolnica-evrope-2016/. Accessed on September 1, 2018.
Mreža FabLab Slovenija. (2018, September). *Mreža FabLab Slovenija.* Retrieved from http://fablab.si/
Naldi, L., Nilsson, P., Westlund, H., & Wixe, S. (2015). What is smart rural development? *Journal of Rural Studies*, 90–101. doi:10.1016.
Razpršeni hotel Konjice. (2018). *Razpršeni hotel Konjice.* Retrieved from https://razprsenihotel.si/
Ringenson, T., Höjer, M., Kramers, A., & Viggedal, A. (2018). Digitalization and environmental aims in municipalities. *Sustainability, 10*(4). doi:10.3390.
Smart Village. (2018). *Smart village: New thinking for off-grid communities worldwide.* Retrieved from https://e4sv.org/about-us/
Stojmenova Duh, E., & Kos, A. (2016). *FabLabs as drivers for open innovation and co-creation to foster rural development.* 2016 International conference on Identification, Infromation and Knowledge in the Internet of Things (IIKI)

(pp. 215–216). Beijing: IEEE. Retrieved from https://doi.org/10.1109/IIKI.2016. 70

SURS. (2017). *Urbano-ruralna tipologija, statistične regije*. Retrieved from https://www.stat.si/dokument/9484/Kartografski%20prikaz%20statisti%C4%8Dnih%20regij%20po%20urbano-ruralni%20tipologiji%202017.pdf. Accessed on August 15, 2018.

SURS. (2018). *Statistični urad, Prebivalstvo*. Retrieved from https://www.stat.si/StatWeb/Field/Index/17. Accessed on November 30, 2018.

UN. (2014, July 10). *World's population increasingly urban with more than half living in urban areas*. Retrieved August 15, 2018, from http://www.un.org/en/development/desa/news/population/world-urbanization-prospects-2014.html

UN. (2018). *Sustainable development goals*. Retrieved from https://www.un.org/sustainabledevelopment/sustainable-development-goals/. Accessed on March 20, 2019.

Visvizi, A., & Lytras, M. D. (2018a). It's not a fad: Smart cities and smart villages research in European and global contexts. *Sustainability*, *10*(8), 2727. doi:10.3390/su10082727

Visvizi, A., & Lytras, M. D. (2018b). Rescaling and refocusing smart cities research: From mega cities to smart villages. *Journal of Science and Technology Policy Management*, *9*(2), 134–145. doi:10.1108/JSTPM-02-2018-0020

Vrbica Šifkovič, S. (2017). *Priročnik za vzpostavitev združnega razpršenega hotela*. Ljubljana: Pravno-informacijski center nevladnih organizacij – PIC. Retrieved from http://www.umanotera.org/wp-content/uploads/2016/04/Priro%C4%8Dnik-za-vzpostavitev-zadru%C5%BEnega-razpr%C5%A1enega-hotela.pdf. Accessed on August 20, 2018.

Zavratnik, V., Kos, A., & Stojmenova Duh, E. (2018). Smart villages: Comprehensive review of initiatives and practices. *Sustainability*, *10*(7). doi:10.3390.

ZGS. (2018). *Gozdnatost in pestrost: Splošni podatki in dejstva o gozdovih v Sloveniji*. Retrieved from http://www.zgs.si/gozdovi_slovenije/o_gozdovih_slovenije/gozdnatost_in_pestrost/index.html. Accessed on August 31, 2018.

Chapter 10

Smart Village Projects in Korea: Rural Tourism, 6th Industrialization, and Smart Farming

Jonghoon Park and Seongwoo Lee

Introduction

To better characterize economic success in South Korea (hereafter Korea), it should be seen as "a prime example of latecomer's high-rate growth, which condenses the longer development history of developed countries" in the world (Cho, 1994, p. 177). National growth was concurrent for growth in agricultural production and incomes. Nonetheless, the agricultural sector in Korea has been lagging behind the manufacturing sector in the last five decades. In particular, urbanization that has led to depopulation of rural areas has accelerated since the mid-1970s, and rural areas have lost its competitiveness and suffered from problems such as an aging population, disappearance of basic industries, and lack of social capital. Further, the farming environment has changed due to the multilateral (e.g., with World Trade Organization) and bilateral (e.g., with the United States) negotiations that result in the liberalization of the agricultural market and lowering of prices of agricultural products. This makes farmers anxious, and previous policies that focused on production increases have reached its limit.

However, rural spaces in Korea are being revitalized. New development strategies have been designed and implemented, which indicates that rural spaces are not going to disappear in Korea anytime soon. Currently, there is an increase in the number of city dwellers moving to rural spaces with regard to quality of life issues such as clean air and decent living environments. In addition, more people are abandoning their jobs in urban areas and turning to farming. The number of people leaving the city to take up farming has jumped from 17,464 in 2011 to 20,559 in 2016 (Statistics Korea, 2018).

However, the growing interest in farming has not been able to stop the steady decline in the number of farmers. While there were 2.96 million farmers in 2011, the number plummeted to 2.5 million by 2016 (OECD, 2018). To respond to the decline, the agricultural policy in Korea has sought to strengthen its competitiveness via such policies as an economy of scale and producing safe and high-quality agricultural products. Since 2005, the government of Korea has made several policy actions targeted to revitalize rural spaces. In addition, there are plans to transform rural villages into spaces for a diverse range of industrial activities as well as a high quality of life. Introducing rural tourism, 6th industrialization, and smart farming to rural spaces are the representative recent streams of the rural development strategies in Korea.

Korea's rural policy has a tendency toward the EU model in that it emphasizes the importance of shift from a top-down strategy to a bottom-up territorial development.[1] As it did in EU countries, the Korean government acknowledges agriculture is not the only driver of rural jobs and wealth creation. Rather it understands that diversified non-farm activities in rural areas are essential to revitalize the rural economy. The purpose of this chapter is to investigate diverse policy experiences of smart village strategy and why rural tourism has been a major development strategy in this stream. This chapter presents the current status and future characteristics of various rural development strategies in Korea. This study also shows the constraints and problems identified during the last two decades based on empirical experiences in Korea.

The argument in this chapter is structured as follows. The section 'Ideologies of Rural Policy' presents ideologies of rural policy based on the theoretical and empirical arguments in the EU countries and the United States. After explaining the rural policy experiences in Korea in the section 'Rural Policies in Korea', the section 'The Smart Village Project in Korea: Rural Tourism' investigates the nature of rural tourism policies in Korea. The section 'Retrospective Evaluations on Rural Tourism Policy in Korea' evaluates the rural tourism policies and section 'Recent Streams: 6th Industrialization and Smart Farming' explains the recent streams of rural development policies in Korea. Finally, section 'Conclusions' concludes with the major findings and implications.

Ideologies of Rural Policy

More than a decade ago, Pezzini (2001) found that rural policy has gone far beyond sectoral policy in most developed countries, a trend that has subsequently intensified in many developed economies. OECD (2009) showed that, for member countries, there is a policy focus on rural issues such as making improvements to rural competitiveness, by increasing human and social capital,

---

[1]The 16th President of South Korea is Roh Moo-hyun, who established the Presidential Committee on Balanced National Development (PCBND). The major purpose of the committee is to construct diverse strategic plans for balanced regional development in Korea. The bottom-up regional development approach was established by the committee incorporating the ideologies of regional development strategies of LEADER in EU.

by developing markets, and by diversifying economic activities. Nonetheless, policy differences exist among member countries.

Poverty is generally higher in rural spaces throughout the world, influenced in large part by demographic and socioeconomic characteristics of both individuals and the prevailing local industrial and labor market conditions. However, Weber, Jensen, Miller, Mosley, and Fisher (2005), based on an extensive literature review, found that the incidence of poverty is higher in rural spaces even after controlling for individual and contextual characteristics. Sanchez-Zamora, Cobos, and Delgado (2014) found that the rural development fund from LEADER (*Liaison entre actions de developement de economie rurale*) is positively correlated to rural well-being in Spain. Markey, Halseth, and Manson (2008) demonstrated that rural decline in northern British Columbia, Canada, has been created by government policy and cannot be attributed to such broader structural changes as urbanization and globalization.

It is true that rural spaces do not evolve homogeneously and feature diverse territorial heterogeneities. Knowing the heterogeneous characteristics of rural spaces is crucial for the execution of diverse policies (Sanchez-Zamora et al., 2014), Bell and Jayne (2010) warn that, if future rural development agendas are to succeed, planners and policymakers must be cognizant of the complexity and diversity of rural spaces. Citing the need for agricultural diversity as a palliative, Marsden and Sonnino (2008) have suggested examining existing and potential governmental roles in pursuit of rural development strategies based on agricultural diversity. Rural spaces differ in a variety of ways, but an integrated and flexible policy response would be more effective. To this end, Rizov (2005) argued that achieving optimal diversity at the community level is one key factor for reducing poverty and promoting economic vitality in rural spaces.

The other side of government subsidies on agricultural policy has to do with the rate of return. Having a grand vision for development and aligning the vision with local planning may create changes that are meaningless unless there is enough funding and support. Esparcia (2014) found that a lack of enough funds restricts consolidation and development. Latruffe, Dupuy, and Desjeux (2013) found that about one-fifth of the farmers who would remain in farming with the Common Agriculture Policy (CAP) program would quit farming if the CAP were removed. The similar effects of the CAP reduction on farmers' nonrural income are presented by Leeuwen and Dekkers (2013).

The financial performance of both agricultural policy and rural policy are worldwide concerns. Recent agricultural policies were expanded to include spatial policies that enhance the environment and quality of life in rural spaces. However, those who make policy often experience skepticism as to the efficacy of policy interventions in rural spaces (World Bank, 2009). Evidence for vanity projects in rural spaces has been found everywhere (Fahrmann & Grajewski, 2013; Woods, 2005, pp. 148–149). Concerns regarding the effectiveness of public expenditures on rural policy apply to the national level as well as subnational levels (Olfert & Partridge, 2010).

Rural Policies in Korea

The economic accomplishment of Korea is an example of high growth rates for straggling countries, that is, solidifying a long development history from developed countries in a shorter time period. However, Korea embraced a lopsided industrial and spatial growth strategy. Any foreseen imbalances in many areas become obvious, such as between urban and rural development, large- and small-scale businesses, and export-oriented or domestic industries. Further, the benefits have been concentrated in just a few areas. Preference has been given to a few predetermined industrial projects concentrated within specific urban areas.

The collapse of rural communities has resulted primarily from urban-focused policies, which are shown with shrinking populations and poor economic conditions. Farm households in 2015 made up approximately 5.4% (~2.6 million individuals) of the total population in Korea, which is considerably less than a quarter of the 1970 levels. Also, in the same time period, a total of 38.4% of rural residents were 65 years old or older; however, the national average was approximately 14.2%. In the economy, agriculture dropped to 2.0% in 2010 from 26% in 1970. Additionally, 26% of the GDP and 50.4% of the total labor force took part in agriculture in 1970; however, in 2015, the percentage of GDP was 2.0% and the total labor force was 5.1%. Rural communities are often unable to achieve economic parity with urban households, with the income gaps between rural and urban households being a concern for policymakers. The average income of rural households was higher (111%) than that of urban households in 1975 with an average rural household income comparable to urban households in the early 1990s. However, rural household income is just 70.4% of urban household income in 2015.[2]

Regional development strategy in Korea generally follows the comprehensive objective of the rural development policy of the EU's CAP. The Korean approach has been highly influenced by the EU experience emphasizing the importance of a bottom-up territorial development. Since the early 2000s, the CAP has been transformed from a sectoral policy of support for agricultural production to a more comprehensive policy that encompasses territorial development and environmental enhancement (Rizov, 2004). Enhancing the quality of life of rural residents and propelling diversification of economic activity in rural spaces are the major goals of the EU's rural development policy, represented by the LEADER (Teilmann, 2012), and the same is true for the rural development policy in Korea (Hwang & Lee, 2015). Improving the quality of life and increasing economic diversity in rural spaces are the primary common points between Korean policy and the EU's rural policies.

A reduction in the economic imbalances for urban and rural spaces is crucial for rural revitalization or to create the market-driven development of rural spaces (Nelson, Oberg, & Nelson, 2010). The impressive growth of the Korean

---

[2]All statistics are from the Korean Statistical Information Service (KOSIS, http://kosis.kr/).

economy, coupled with its increased integration into world markets, has made agricultural policy an issue in diverse bilateral and multilateral international trade negotiations. To sustain rural resilience,[3] there have been many local-, provincial-, and central-level policies in Korea that address economic and population disparities between urban and rural spaces since the Uruguay Round Agreement in 1992. To decrease disparity, the government developed rural policies in the early 1990s when the economic imbalance between rural and urban areas was increasing.

Agricultural policy in Korea until the 1990s was the same as rural policy, which was also the same as Organization for Economic Cooperation and Development (OECD) countries. In the 1970s and the 1980s, policies focused on enhancing productivity for agricultural products with a focus on self-sufficiency for rice. In the 1990s, structural adjustments have been prepared to increase competitiveness. However, Burmeister (1992) showed that the shift in agricultural policy hindered rural development and subsequently the economic viability of rural households. The Korean government acknowledges agriculture is not the only driver of rural jobs and wealth creation. Rather it understands diversified non-farm activities in rural areas are essential to revitalize the rural economy.

The government of Korea floated a plan for agriculture and rural communities, which also established a mid- and long-term policy framework (cf. Table 1). Huge investments were made to satisfy specific goals, with 42 trillion Korean Won (47 billion USD) and 45 trillion Korean Won (37 billion USD) invested in the first and second phases of the plan. The third phase will invest 119 trillion Korean Won (104 billion USD). The expenditures increased the budget which was set aside for agriculture to 13–15% during the 1994–2013 fiscal periods, that is, an increase from 9% in 1993.[4]

The major policies in the revised fourth national territorial plan (2000 to 2020) relevant to the development of rural smart village are, first, establishing a regional innovation system fit for depressed regions, second, inducing agriculture to become value-added industries, third, diversifying rural economic activities and integrating industrial support, fourth, improving the welfare of rural residents by improving settlement conditions, and finally, encouraging rural–urban interaction.

The Smart Village Project in Korea: Rural Tourism

There has been a shift in agricultural and rural policy paradigms that is mirrored by a shift from productivism to post-productivism. Productivism emphasizes

---

[3]Resilience has many meanings in the contexts of agricultural and rural development (McManus et al., 2012, pp. 21–22). Here, the term mainly implies economic resilience: the capability to embrace changes from diverse external shocks and to sustain its status quo.
[4]Farmers in Korea have a large influence on policymakers in the legislative and the executive branches of central government, despite their small numbers.

Table 1. Agricultural and Rural Investment Plans in Korea (1991 to 2017).

Phases	Titles	Period	Major Objective	Budget
1st	Agricultural and Rural Structure Improvement Plan	1991–1996	– Strengthening the competitiveness of the agricultural sector	42 trillion Korean Won (47 billion USD)
2nd	Agricultural and Rural Structure Adjustment Plan	1997–2003	– Strengthening the competitiveness of the agricultural sector and rural areas	45 trillion Korean Won (37 billion USD)
3rd	Comprehensive Plan on Agriculture and Rural Communities	2004–2017	– Development of agri-food sector – Enhancing the competitiveness of the agricultural sector – Enhancing the quality of life in rural areas	119 trillion Korean Won (104 billion USD)

Source: Hwang, Park, and Lee (2018).

intensity and productivity for farming. Conversely, post-productivism requires horizontally integrated rural communities. Many developed countries also concentered on horizontally integrated rural communities (Woods, 2005). Various rural policies were planned and implemented in the early 1990s when there was a significant gap between rural and urban spaces. Rural tourism is considered a preferred policy to resolve the problem; however, most programs operated by diverse ministries were duplicated or scattered at that time because each ministry planned and conducted its own strategy independently.

Rural tourism has been embraced by many countries as a desirable rural policy to create rural viability (Brandth & Haugen, 2011; Getz & Page, 1997; Kannan & Singh, 2006; Liu, 2006; Sharpley & Vass, 2006). One of the most significant elements of rural changes has been a transition from an economy based on production to an economy based on consumption (Woods, 2005) Rural tourism is defined as the combined utilization of agricultural, eco-, and cultural products that encompass a diverse set of economic, social, educational, environmental, recreational, or therapeutic functions, which have been less utilized before. Rural tourism allows for the rediscovery of rural resources that have been disregarded due to modernization and urbanization that provide awareness to farmers and policymakers to embrace a larger perspective than a traditional, agricultural product-oriented mindset for rural development (Lee & Kim, 2010).

Rural tourism is diverse, but it includes agricultural production, lifestyles, and rural spaces. Policymakers and rural dwellers see this as a viable path to reinvigorate rural economies. The government expects to increase the quality of life with a new development approach for rural spaces. Douglass (2000) noted that if an approach like this is to work well, then it is important to develop rural—urban linkages as they could be a driver for a diverse and multiplier effect. The government assumes that diversification in a rural economy needs to be stimulated to harmonize the promotion of primary products as well as the tourism industry.

Since the campaign of smart rural village as a rural development strategy is closely related with the discussion of rural tourism in Korea, this study investigates past and recent streams of rural tourism strategies pursued by the central government in Korea. Currently, rural tourism is an area that Korean farmers could benefit from to offset the declining agricultural incomes. The government put herculean efforts into this new program since 2003, that is, the same year the government proposed agriculture and rural plans to spend about 119 trillion Korean Won (about 119 billion USD) from 2004—2013.

The government promulgates a variety of urban—rural exchange programs and tourism activities. Table 2 shows rural tourism programs that have mainly focused on bringing city people to rural villages. Urban—rural exchange programs are opportunities for both rural and city people. City people can partake in activities and learn by experience, staying and resting in the countryside, whereas it results in rural people having jobs and increasing sales. The major outputs of these programs are as follows: first, it fosters tourism professionals to take the lead in rural tourism, second, it intensifies public relation (PR) and marketing activity to create needs for rural tourism, third, it develops village conditions to satisfy city dwellers' needs, and finally, it creates an institutional arrangement for activation of urban—rural exchange.

In addition, the MAF adopted a comprehensive agenda and the development of rural tourism is a major policy alternative. The plan was the first comprehensive rural plan, which is titled the Comprehensive Rural Village Development Program (CRVDP). The plan expects to build 1,000 rural zones that combine three to five villages in each zone. Approximately five to seven rural development zone per city or county that have rural characteristics had been developed by 2013. The total budget for this plan was seven billion Korean Won (about 7 million US dollar) per rural development zone.

Since 2010, the major policy shift from the provision of hardware-oriented policy to the software-oriented policy was made with the advance of the post-productivist paradigm in agriculture and rural spaces in Korea. In addition, the administrative entity to handle regional and rural development policy has been changed from central government to local government. The new initiative is called a block grant system on which the whole accounting of the regional development grant is operated and managed by the PCBND. In this system, broader sets of goals are set at the central government level and are interpreted and implemented at a lower level and typically by the second-level local autonomy,

Table 2. Rural Tourism Programs in Korea in the Post-productivist Era.

Project Name	Information Network Village	Green Tour Village	Fishing Experience Village	Rural Theme Village	Mountain Theme Village
Implementing Agency	Ministry of Government Administration and Home Affairs	Ministry of Agriculture and Forestry (MAF)	Ministry of Maritime Affairs and Fisheries	Rural Development Administration	Korea Forest Service
Objectives	• To promote regional informatization development in rural communities • To increase profitability of rural communities through informatization	• To provide farming and rural community tour for city dwellers • To increase off-farm income of farming communities	• To provide hands-on fishing program through establishing fishing tourism facilities • To increase off-fishing income	• To improve rural amenity through traditional rural knowledge and culture	• To utilize forestry resources for additional income • To achieve balanced national development by improving landscape and living conditions of forestry areas
Action Plan	• Reduce the gap of information • Revitalize regional economy through informatization	• Establish common facilities for green tour program	• Improve roads, parking facilities, lights, etc. • Establish information center,	• Set up own characteristics of village • Establish facilities for hands-on program	• Establish a system for increase of forestry products • Using unique propensity of forestry and develop forestry tour program

Smart Village Projects in Korea

	• Facilitate green tour program	• Improve landscape of participating villages	squares, camping lot, etc.	• Program development by experts
Period	2001–2012	2002–2013	2002–2009	2002–2009
Unit	Village	Village	Village/Municipality	Village
No. of Villages	280(39)	123	31	66
Cost/Unit*	5 (central govt's support 3.5)	2 (govt's support 50%)	5 (govt's support 50%)	2 (govt's support 50%)
Duration	1 year	1 year	1 year	2 years
Total Cost*	1,052	246	359	112

Period	1995–2007
Unit	Village
No. of Villages	153
Cost/Unit*	14
Duration	3 years
Total Cost*	2,413

Source: The Office for Government Policy Coordination Assessment of Green Tour Village Programs.
Notes: *100 million Korean Won, about 105,000 USD.

that is, city/county. Thus, most rural tourism programs are now independently planned and managed by local government with major financial support (50% to 80% by program characteristics) from the block grant system.

Retrospective Evaluations on Rural Tourism Policy in Korea

Rural theme village programs promulgated by each related ministry have shown similarities in the content and objectives (Table 2). While the standard of living is consistently improved for those rural villages participating in the program (Hwang & Lee, 2015), Lee and Nam (2006) argued that much is required to reach a certain level of stable rural community. The major problems are summarized as follows.

- Conflicts due to the similarity of program contents
  Despite analogous geographical and physical conditions, there are ample discrepancies in subsidies for participating villages in similar programs dependent upon government ministries. Particularly, the contents of MAF and the Rural Development Administration are quite similar, which causes confusion as well as dissatisfaction for those participating farmers.
- Post-hoc management after implementation is not apparent
  Despite educational programs that bolster human capital development for participating farmers to design and stage implementation, no government ministries are providing management programs after a program has been implemented, which brings difficulties for sustaining the program's objectives.
- Sustainability of the participating villages
  There have been no increases in the population for villages that participated in the programs. Further, these programs are less successful in increasing rural incomes. It may be caused by poor-quality content or similarities among the programs that failed to attract city dwellers.
- Growing dependency on government subsidies
  Rural tourism programs hope that villages are able to achieve independent management and sustainable growth after the program was completed. However, the participating villages become more and more dependent on government support, even in cases where the project's implementation was seen as successful by the MAF.

Recent Streams: 6th Industrialization and Smart Farming

Since 2015, the government of Korea started two new programs that promote tangible products that typify the rural community. These two programs are 6th industrialization and smart farming. Incorporating comprehensive concepts of business ecosystem and new information technologies in agriculture and rural areas, these programs intend to restructure the rural economy and promulgate its potential by contributing to rural economic sustainability.

6th Industrialization

The government tried to change agriculture into a 6th industry, that is, agriculture is integrated with manufacturing, processing, marketing, and tourism services to create new added-value products (OECD, 2018). Each Korean province established and operates a 6th industry support center to invigorate rural communities (Lee & Hwang, 2016).

In the workplan for 2016, the Ministry of Agriculture, Food and Rural Affairs presented measures to boost rural economy and exports by transferring agriculture into a sixth industry, which includes linking production and exports to tourism. The 6th industrialization of agriculture is a strategy for integrating agricultural production with processing and sale, and forming a business ecosystem which includes tourism or interactions to create jobs and added values (KREI, PRIMAFF, & IAED, 2014). In plain language, the 6th industrialization attempts to internalize jobs and added values related to the food industry, tourism, and service business which have gone out of agriculture (Figure 1).

The 6th industrialization is divided by the following types of development entities dependent upon objectivities, regional conditions, management type, leading industries, and methods of cooperation. The types are classified as following: community type (local community), franchise type (agreement and transaction), and network types (linkage between agriculture, industry, and commerce).

The 6th industrialization affects the rural area in diverse ways such as job creation, producing and consuming safe agricultural product and food, creating new values, and recovering local economy. Today, the 6th industrialization receives a lot of attention because it is a more comprehensive approach than previous policies.

Smart Farming

Smart farm is a new technology that combines agriculture with information and communications technology (ICT) and enables year-round production through

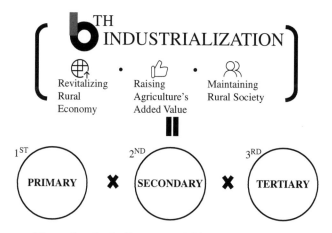

Figure 1. Basic Concepts of 6th Industrialization.

improving productivity and quality of products by modernizing and connecting farms. Korea desperately needs to adopt the smart rural system and combine it with its most developed ICT technology to grow its agricultural industry because the industry is being weakened by the aging society and small-scale farming.

The concept of smart farming is still considered relatively new in Korea, where most farmers come from an older generation that is generally less willing to adopt a completely new approach to growing and maintaining crops. The government of Korea supports the introduction of smart farming technology on 4,000 hectares of land. While smart farming is yet to find a home at most farms in Korea, it has already been put into practice in the world. Netherlands, Israel, Denmark, Germany, and Japan have already started using ICT to improve its effectiveness as well as to increase productivity.

The Ministry of Agriculture, Food and Rural Affairs (MAFRA) revealed policies to attract young workers to smart farming by providing training courses and building large-scale smart farm complexes in 2018. Smart farming brings information and communications technology to greenhouses and livestock farms and allows farmers to remotely control the growing environment of crops and animals through smartphones and computers.

The latest announcement focuses on creating innovation models, a shift from the previous plan to introduce smart farming at existing farms, under which young entrepreneurs and farmers will be trained and other related businesses will be supported. Training courses will be available in 2019 and will last for 20 months, with plans to create 600 professionals by 2022. In addition, young farmers who complete the courses will be given government subsidies to open a smart farming business on 30 hectares of land that will be allotted to them, while agricultural loans at lower interest rates will be available for both new and experienced farmers.

Plans to build a smart rural test site are underway where research projects, exhibitions, and tests will take place to help enhance the competitiveness of smart farms. A total of four smart farming innovation valleys will be built across the country by 2022, along with multipurpose venues consisting of a training center, smart farms for rent, and a test site. The MAFRA is hopeful that the move will help create more than 4,300 jobs over the next few years in the burgeoning smart farming industry.

Conclusions

This chapter investigated diverse policy experiences of smart village strategy in Korea. Along with introducing the historical development strategy in Korea, this study presented the current and possible future characteristics of rural development strategies in Korea. This study investigated the perceived role of tourism as well as recent streams of rural development policies such as 6th industrialization and smart farming in the rural development strategies. Presenting success and failure stories, this study also considered why development of rural tourism has been slow in rural areas in Korea, reviewing restraints, reservations, and problems identified during the last decades in Korea.

In assessing the managerial aspects of rural tourism policies that were carried out by diverse government agencies, it seems that the program has not reached its policy objectives, and may even be gone to an undesirable direction. Many of the tourism farms are not yet profitable, and some of them are in a state of financial crisis. It also seems that their contribution to the local economy is not as great as expected.

However, in a rare empirical evaluation on the rural tourism policy, Hwang and Lee (2015) investigated a rural tourism program for the clarification of policy implications in the past performance as well as in the future direction. They found a positive effect of rural theme village program on non-farm income for participating farmers. To increase the economic efficiency of the program, they suggested that government support should be directed to activities in rural areas with public good nature such as promotion, training, network establishment, and information system network, rather than direct financial provision for individual farmers.

Based on the experiences of the rural tourism programs in Korea, this study proposes the following policy implications.

First, development or planning has an inherent characteristic of future orientation. The government needs to introduce market-friendly policies such as reinforcement of liaison between rural and urban areas for inflow of urban capital into rural areas in the long-term perspective. The principle that works in a market is to concentrate limited available resources on potential opportunity factors maximizing efficiency rather than to compensate for a drawback. The problem is that rural development strategy in Korea relied heavily on a top-down development strategy by public sector and resulted in numerous vanity projects. It would be much more helpful to adopt market-friendly policies such as reinforcement of liaison between rural and urban areas for inflow of urban capital into rural area. The fundamental device to avert rural Korea from its deepening emptiness and relative disparity in income is to foster a market-friendly environment thorough rural development policy raising the competitiveness of rural Korea.

Second, it also may be necessary to have a systematic program of design, monitoring, and follow-up by the development agencies concerned, including agricultural cooperatives and the related ministries. To achieve this, the construction of diverse levels of public–private partnership in the rural development strategies is essential. Without suitable cooperation of public–private partnership in carrying out rural tourism program, there can be a lack of success for rural tourism entities if the public sector solely wields the whole processes of the program.

Third, the local government should take a leading role to form a regional partnership for the rural–urban exchange programs. Most of all, the government should have a long-term perspective in terms of sustainable agriculture and rural development since we have to expect a bunch of pitfalls in the development of the rural tourism as a strategy to revitalize our rural economy during the last decades.

Finally, few government agencies provide management programs after the implementation of the program. This makes it difficult for participating farmers to sustain the program. Government should prepare supplementary programs for farm households to guarantee continuous successful management of the program.

References

Bell, D., & Jayne, M. (2010). The creative countryside: Policy and practice in the UK rural cultural economy. *Journal of Rural Studies, 26*(3), 209–218.

Brandth, B., & Haugen, M. S. (2011). Farm diversification into tourism-implications for social identity? *Journal of Rural Studies, 27*(1), 35–44.

Burmeister, L. L. (1992). Korean minifarm agriculture: From articulation to disarticulation. *The Journal of Developing Areas, 26*(2), 145–168.

Cho, S. (1994). *The dynamics of Korean economic development.* Seoul: Institute for International Economics.

Douglass, M. (2000). *Turning points in the Korean space-economy: From the developmental state to the intercity competition, 1953–2000.* The *Urban Dynamics of East Asia, Discussion Papers.* Asia/Pacific Research Center, Institute for Studies, Stanford University.

Esparcia, J. (2014). Innovation and networks in rural areas. An analysis from European innovative projects. *Journal of Rural Studies, 34*, 1–14.

Fahrmann, B., & Grajewski, R. (2013). How expensive is the implementation of rural development programmes? *European Review of Agricultural Economics, 40*, 541–572.

Getz, D., & Page, S. J. (1997). Conclusions and implications for rural business development. In S. Page & D. Getz (Eds.), *The business of rural tourism: International perspectives* (pp. 191–205). London: Cengage Learning EMEA.

Hwang, J., & Lee, S. (2015). The effect of the rural tourism policy on non-farm income in South Korea. *Tourism Management, 46*, 501–513.

Hwang, J., Park, J., & Lee, S. (2018). The impact of the comprehensive rural village development program on rural sustainability in Korea. *Sustainability, 10*, 2436, doi:10.3390/su10072436

Kannan, D., & Singh, V. (2006). Management for successful agro-tourism in India. Paper presented at the Asian Productivity Organization Seminar, February 20–27, Taiwan, Republic of China.

KREI, PRIMAFF, & IAED. (2014). *The 6th industrialization of agriculture.* Seoul, Korea: Korea Rural Economic Institute.

Latruffe, L., Dupuy, A., & Desjeux, Y. (2013). What would farmers' strategies be in a no-CAP situation? An illustration from two regions in France. *Journal of Rural Studies, 32*, 10–25.

Lee, S. W., & Hwang, J. H. (2016). The 6th industrialization in post-productivist era in Korea. Korea–Japan Rural Planning Seminar, Kanazawa, Japan.

Lee, S. W., & Kim, H. J. (2010). Agricultural transition and rural tourism in Korea: Experiences of the last forty years. In G. B. Thapa, P. K. Viswanathan, J. K. Routray, & M. M. Ahmad (Eds.), *Agricultural transition in Asia* (pp. 37–64). Bangkok: Asian Institute of Technology.

Lee, S. W., & Nam, S. Y. (2006). Agro-tourism as a rural development strategy in Korea. *Journal of Rural Development, 29*(6), 67–83.

Leeuwen, E. V., & Dekkers, J. (2013). Determinants of off-farm income and its local patterns. A spatial microsimulation of Dutch farmers. *Journal of Rural Studies*, *31*, 55–66.

Liu, A. (2006). Tourism in rural areas: Kedah, Malaysia. *Tourism Management*, *27*(5), 878–889.

Markey, S., Halseth, G., & Manson, D. (2008). Challenging the inevitability of rural decline: Advancing the policy of place in northern British Columbia. *Journal of Rural Studies*, *24*(4), 409–421.

Marsden, T., & Sonnino, R. (2008). Rural development and the regional state: Denying multifunctional agriculture in the UK. *Journal of Rural Studies*, *24*(4), 422–431.

McManus, P., Walmsley, J., Argent, N., Baum, S., Bourke, L., Martin, J., Pritchard, B., & Sorensen, T. (2012). Rural community and rural resilience: What is important to farmers in keeping their country towns alive? *Journal of Rural Studies*, *28*, 20–29.

Nelson, P. B., Oberg, A., & Nelson, L. (2010). Rural gentrification and linked migration in the United States. *Journal of Rural Studies*, *26*, 343–352.

OECD. (2009). *Rural policy reviews*. Paris: OECD.

OECD. (2018). *Agricultural policy monitoring and evaluation*. Paris: OECD.

Olfert, M. R., & Partridge, M. D. (2010). Best practices in twenty-first-century rural development and policy. *Growth and Change*, *41*, 147–164.

Pezzini, M. (2001). Rural policy lessons from OECD countries. *International Regional Science Review*, *24*(1), 134–145.

Rizov, M. (2004). Rural development and welfare implication of CAP reforms. *Journal of Policy Modeling*, *26*(2), 209–222.

Rizov, M. (2005). Rural development under the European CAP: The role of diversity. *The Social Science Journal*, *42*(4), 621–628.

Sanchez-Zamora, P., Cobos, R. G., & Delgado, F. C. (2014). Rural areas face the economic crisis: Analyzing the determinants of successful territorial dynamics. *Journal of Rural Studies*, *35*, 11–25.

Sharpley, R., & Vass, A. (2006). Tourism, farming and diversification: An attitudinal study. *Tourism Management*, *27*(5), 1040–1052.

Statistics Korea. (2018). Korean Statistical Information Service (KOSIS). Retrieved from http://kosis.kr/.

Teilmann, K. (2012). Measuring social capital accumulation in rural development. *Journal of Rural Studies*, *28*(4), 458–465.

Weber, B., Jensen, L., Miller, K., Mosley, J., & Fisher, M., 2005. A critical review of rural poverty literature: Is there truly rural effect? *International Regional Science Review*, *28*(4), 381–414.

Woods, M. (2005). *Rural geography*. Thousand Oaks, CA: Sage.

World Bank. (2009). *World Development Report: Reshaping economic geography*. Washington, DC: The World Bank.

Chapter 11

Smart Villages and the GCC Countries: Policies, Strategies, and Implications

Tayeb Brahimi and Benaouda Bensaid

Introduction

Today, members of the Gulf Cooperation Council (GCC), namely, Kingdom of Saudi Arabia (KSA), Qatar, the United Arab Emirates (UAE), Kuwait, Bahrain, and Oman, have in recent years witnessed transformation on an unprecedented scale with regards to shift to smart cities and smart villages (Deloitte, 2017). According to the International Telecommunication Union (ITU) Report (Wafula, 2016), 10 out of 22 Arab countries have been engaged in Sustainable Smart Cities (SSC) using the Brownfield model where development is built upon existing cities like Barcelona and Singapore, or the Greenfield models where cities with heavy urbanization are specially designed and constructed originally to be smart such as those cities in the Middle East, India, and China (Webb, 2015). In fact, urban planners in most GCC cities are already including smart city infrastructure and features into their new transformation plans and visions (Deloitte, 2017; McKinsey, 2018). Investment in smart cities is expected to grow in most countries of the GCC, with an increased focus on KSA and UAE. Currently, the GCC countries are building on international best practices and leveraging opportunities to transform into digital government models with key themes articulated around smart cities, smart tourism, next-generation care, classroom of the future, smart government, and future of mobility.

The objective of this chapter is to examine the experience of the GCC countries in developing smart villages, especially with reference to their new vision on revitalizing rural areas. This research reviews the current approaches, strategies, and initiatives taken toward achieving this goal, alongside highlighting the potential opportunities, problems, and challenges ahead of the vision. We will first draw on the transformational context of the GCC countries with respect to smartizing villages, then proceed with the development of smart cities and how each GCC country is currently weaving its plans toward smart villages. This study addresses several key aspects of both the theory and practice of smart village models in each of the GCC countries. In addition, it addresses the contextual policy model frameworks and the potential challenges and barriers of smart village project implementations. It also provides recommendations and solutions based on current government initiatives and pilot projects. To this end, first, we will outline the concepts of smart cities and smart villages. Then the GCC context, and the ongoing National Transformation Plans, are discussed. Then some world initiatives of smart cities and smart villages are reviewed. Finally, smart village initiatives in the region of the GCC are elaborated. Conclusions and recommendations follow.

The Concept of Smart Cities and Smart Villages

Multiple definitions of smart cities already exist (Albino, Berardi, & Dangelico, 2015; Mora, Bolici, & Deakin, 2017; Visvizi & Lytras, 2018a). None, however, can be universally adopted. The ITU (ITU, 2014) listed 100 definitions of smart cities. Definitions appear to be lacking explicit connection to the creation of a better environment for all the citizens (Lara, Costa, Furlani, & Yigitcanlar, 2016). Some definitions are found to be technology-oriented while others rather focus on the social impact. Lytras and Visvizi (2018) examined the debate surrounding smart cities according to the complex perspective of citizens' awareness and their ability to use smart services. They called for a careful rethinking of the focus and rationale underlying the debate on smart cities (Visvizi & Lytras, 2018b). Smart cities, however, appear to share salient features like (1) capitalizing on information and communication technology (ICT); (2) promoting new technologies and sustainable development; (3) investing in human capital; (4) integrating urban infrastructure, culture, and services; (5) enhancing the competitiveness of cities, and improving the quality of life of citizens; (6) promoting new forms of governance, public participation, and social inclusion; (7) reducing ecological footprint while seeking renewable energy resources; and (8) promoting creative economy and knowledge-based society (Albino et al., 2015).

The concept of the smart village is still new. However, its thrust is about the collection and strength of each community member alongside the integration of ICT for the benefit of the rural community (Visvizi & Lytras, 2018a). According to Ankur, Shyamu, and Gaurav (2018), a smart village essentially should include food security for all, health welfare, environmental development, personality development, democratic engagement of people, efficient governance, social development, education, local business, and use of renewable energy, among others. At the conceptual and empirical levels, Visvizi and Lytras (2018a) argue that the smart village has the potential to bypass the weaknesses of smart cities. In a European context of definition (European Commission, 2017), smart villages refer to "rural areas and communities which build on their existing strengths and assets as well as on developing new opportunities" where "traditional and new networks and services are enhanced by means of digital, telecommunication technologies, innovations and the better use of knowledge." Moreover, using public services efficiently and reducing the impact on the environment can be added to the support of digital technologies that help in improving the quality of life. It is also recognized that these technologies are key to building strong and efficient governance and citizen involvements (Lytras & Visvizi, 2018; Zavratnik, Kos, & Duh, 2018). This brings to light the contribution of the 'Smart Village Network' set to connect villages and associations across Europe and to help them exchange information and experiences (Visvizi & Lytras, 2018a).

The UN's 'Transforming Our World: The 2030 Agenda for Sustainable Development' (UN, 2015) does not explicitly discuss rural areas or villages. However, the second goal of sustainable development shows interest in increasing investment in rural infrastructure, agricultural research and extension services, technology development, and plant and livestock gene banks in order to enhance agricultural productive capacity in developing countries. Least developed countries, in particular (UN, 2015, p. 13). Likewise, item 11 on the list of the goal of Sustainable Development Goals (SDGs) reads: "Support positive economic, social and environmental links between urban, peri-urban and rural areas by strengthening national and regional development planning" (UN, 2015, p. 19). UN-Habitat (UN, 2017) also focuses on building a brighter future for villages, towns, and cities of all sizes. The UN-Habitat report (UN, 2017) reaffirms that to be sustainable, both rural and urban areas need to be developed together in all the domains of economy, environment, and society alongside bridging the existing gap. Here, the concept of urban–rural linkages is based on complementing urban and rural areas with focused attention on the spatial flow between urban and rural areas, mobility, urbanization, food security, and sustainability chain, environment, governance, partnerships, and inclusive investment.

In view of the increasing interest in smart villages, it is worthwhile to address key relevant initiatives to better understand the relationship between theory and practice and also highlight some of the main challenges lying ahead of smart village implementation. Several initiatives use or promote the concept of smart villages (Zavratnik et al., 2018). The *Smart Village Initiative* is set to "ensure access to affordable, reliable, sustainable and modern energy for all" (UN, 2015) by encouraging the use of renewable and clean energy which also requires extensive use of ICT (Gevelt et al., 2018; UN, 2010). The initiative sought to identify and provide energy services to rural areas: East Africa, West Africa, South Asia, Southeast Asia, South America, Central America, the Caribbean, and Mexico. It is essentially committed to linking universal renewable energy access to the development of rural areas such as healthcare, education, clean water, and economic growth since access to energy represents the main driver of sustainable development in rural areas. In a recent study on the link between renewable energy and rural development, the European Court of Auditors (ECA, 2018) concluded that there are high potential synergies and rural development, but those synergies remain largely unrealized as yet.

The *EU Action for Smart Villages and Reflections on Villages of the Future* is set to revitalize rural local communities, promote sustainable development and innovation, and make them more attractive through mobilizing ICTs while resolving the problems of poverty and depopulated rural areas (European

Commission, 2017). However, the Institute of Electrical and Electronics Engineers (IEEE) Smart Village: Empowering Off-grid Communities (IEEE, 2018) seeks to help develop rural areas, off-grid communities, and to further facilitate access to affordable, reliable, sustainable, and modern energy for all (UN-SDG, 2018). This initiative establishes off-grid communities and conducts community-feasibility surveys for electricity, clean water, healthcare, education and learning, and agriculture. It currently serves many rural areas located in Cameroon, Haiti, India, Kenya, Nigeria, and South Sudan.

The International Crops Research Institute for the Semi-Arid Tropics (ICRISAT, 2016) was launched to assist and help farmers to adapt to climate change in climate-smart villages using watershed management to practice climate-smart agricultural, simulated future scenarios up to 2050, digital technologies for smart agriculture, metrological advisory and farm systems approach, and climate and crop modeling to help farmers in the drought areas. The Consultative Group for International Agricultural Research (CGIAR) Research Program on Climate Change, Agriculture, and Food Security (CCAFS; CGIAR, 2017) is another initiative promoting 'climate-smart villages' and adopting less risk to climate change and global warming through smart agriculture with a sustainable increase of productivity and income while building resilience to climate change (Zavratnik et al., 2018). This program is being put into practice in different areas such as West and East Africa, Latin America, South, and Southeast Asia. Regions in the GCC may benefit from such a program, in particular in climate-smart technologies, such as solar irrigation, and climate information services for agricultural planning and risk management.

'A Better Life in Rural Areas' started in 2016, with the proposal of Cork Declaration (European Union, 2016) concerns rural areas, specifically rural exodus and youth drain. It sees investment in rural areas as a necessary, especially with regard to encouraging their identification processes, acknowledging their potentials for economic growth, and ensuring that they will become attractive places for people of all ages to live in and work at. The 'Bled Declaration' (Bled, 2018) is yet another rural digital economy program launched to improve the life quality of rural citizens and tackling the current depopulation of and the migration from rural areas. The Declaration seeks to reach better conditions for developing farming enterprises and a new service sector. Far from Europe, and in the context of India, smart villages focus rather on holistic rural development, derived from Mahatma Gandhi's vision which goes as follows:

> The best, quickest and most efficient way is to build up from the bottom. Every village has to become a self-sufficient republic. This does not require brave resolutions. It requires brave, corporate, intelligent work. (Ramachandra, Hegde, Subhash Chandran, Kumar, & Swamiji, 2015)

According to the philosophy of Gandhi, smart villages need to provide global means to local needs (Ankur, et al., 2018).

Depending on its respective country or region of development, the concept of the smart village remains subjective. The idea builds upon existing capital/assets alongside developing new ones. It mostly relies on creating a work-life balance for individuals and the insertion of innovative digital technology solely where it is needed for the support of the rural community development in a smarter and greener way (Hogan, Crețu, & Bulc, 2016, European Commission, 2017). Visvizi and Lytras (2018a) highlighted the short-, medium-, and long-term challenges and the corresponding urgency of action and how smart city research can fertilize smart village research. A closer look at the European practices shows that global initiatives focus on different socioeconomic conditions and the issue of sustainability and ICT integration and accordingly propose different solutions adapted to the needs of their respective communities. Sometimes, those global initiatives focus on specific areas with a lack of basic infrastructure (electricity, water supply, internet access, etc.).

The GCC Transformational Context

Despite the abundant natural resources, having one of the largest oil producers (KSA) and the second largest gas producer (Qatar), and most certainly positive economic forecast, the GCC region currently shares multitudes of challenges and risks: population growth, rapid urbanization, phenomenal movement from rural to urban areas, crowded cities, increase in employment demands, securitization and good governance, and shortages of energy, food, and water (Abdulla, 2015; Al-Ebrahim, 2017; Ramady, 2012; The Economist, 2010). Although the GCC is one of the wealthier regions in terms of gross domestic product (GDP)/capita, it has one of the fastest-growing population with an actual total population of 56.4 million (2018), but forecasted to reach nearly 63 million in 2025 and more than 66 million in 2030 as shown by Figure 1 based on the estimates and projections of the urban and rural populations data published by the Population Division of the Department of Economic and Social Affairs (DESA) of the United Nations (UN, 2018).

These statistical forecasts need to be examined in the light of the United Nation Report (UN, 2018), according to which in 2018, 82% of the population in North America live in the urban area followed by 74% of the population in Europe, while in Africa only 43% of the population lives in urban areas. In Asia, it has been reported that the level of urbanization reaches 50%. Moreover, the global rural population is growing slowly; there are about 3.4 billion people living in rural areas, with Africa and Asia alone having up to 90% of the rural population of the world. By 2050 global urbanization is expected to reach 34% in rural areas and 66% in urban areas, roughly the opposite to the global rural—urban population in the mid-twentieth century.

The data published by the United Nations' (DESA)/Population Division (UN, 2018) lists all population by country, covers the period of 1950—2050, and includes the total population as well as the percentage of the population in urban and rural areas. The extracted data for the six GCC countries for the

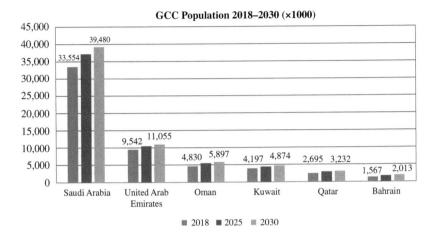

Figure 1. GCC Population Distribution and Estimation (2018–2030). *Source*: Data from UN (2018).

Table 1. GCC Population, Urban and Rural Areas.

	The Population of Urban and Rural Areas (in Thousands) in 2018 and 2030 (UN, 2018)					
	Urban 2018	Rural 2018	Percentage Urban	Urban 2030	Rural 2030	Percentage Urban
Saudi Arabia	28,133	5,421	84	34,143	5,448	86
UAE	8,256	1,286	87	9,865	1,228	89
Oman	4,083	747	85	5,407	582	90
Bahrain	1,399	168	89	1,828	184	91
Qatar	2,672	23	99	3,217	18	99
Kuwait	4,197	0	100	4,874	0	100

years 2018–2030 is shown in Table 1. Figure 2 shows that in the GCC, the majority of the population lives in urban areas. However, this number according to Figure 3 does not decrease much in 2025 and even 2030. In fact, in 2030, the percentage of population living in rural areas will only be 14% in the KSA, 11% in the UAE, 10% in Oman, 9% in Bahrain, 1% in Qatar, and 0% in Kuwait. Presently in the KSA, the largest country in the GCC, 84% of citizens live in urban centers. The government is aware that there is a disparity in the preparedness of its cities to shift to smart cities and, hence, is committing itself to develop many upcoming smart cities and modernizing infrastructure across its municipalities, including its major initiatives. As an example of the municipal

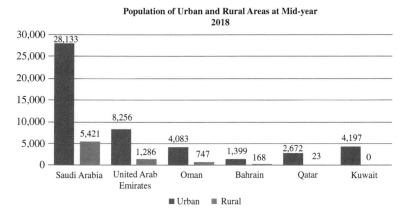

Figure 2. GCC Population, Urban and Rural, in 2018. *Source*: Data from UN (2018).

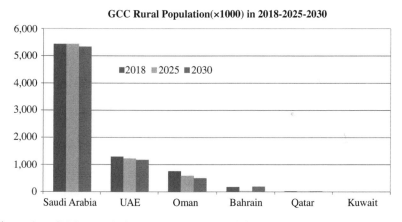

Figure 3. GCC Population, Urban and Rural, in 2018, 2025, and 2030 *Source*: Data from UN (2018).

transformation projects emanating from the Saudi National Transformation Program 2020 and Vision 2030, the Ministry of Municipal and Rural Affairs launched the application of smart city concepts (Saudi Press, 2017).

Although representing only 16% of the population in KSA, rural areas bring with them series of needs like developing rural communities, creating jobs, improving healthcare, education, services, transportation, access to sustainable energy, and boosting farm productivity. With the vast majority under the age of 25 reaching 34%, 46%, and 50% in Kuwait, Saudi Arabia, and the United Arab Emirates, respectively (UN, 2018), this youth bulge represents a golden asset to the GCC nations to move forward toward sustainable economic development and well-being. However, youth capital comes with its own challenges and problems, which need to be harnessed with respect to further developing of both smart cities and smart villages. In a survey conducted by Booz & Company's Ideation Center (AlMunajjed & Sabbagh, 2011), it has been found that in KSA, UAE, and Qatar, youths are concerned about five critical issues, namely education, employment, gender gap, leisure activities, and community engagement. The report states that 87% of youth consider unemployment a major problem, 32% think their education system has not prepared them for the job market, 28% are active in community development, 88% consider surfing the internet as their most popular leisure activities, and 65% want their governments to develop youth service programs and create more opportunities for the young generation. Studies on major challenges affecting the lives of citizens show the following results: high cost of living (62%), finding a suitable job on completion of education (55%), unemployment (45%), getting good quality education (31%), lack of opportunities to express oneself (28%), finding decent housing (27%), financial support of family (27%), family care (20%), weakening of local tradition values (15%), and (9%) for none.

The GCC governments have currently embarked on National Transformation Plans toward a digital transformation toward sustainable development. However, an unmanaged growth is not only critical but could also bring about negative side effects. Aware of this challenge, the GCC region has recently witnessed a growing shift in favor of digital transformation in the economy, diversity of income sources, sustainability, technology, public and world-class government services, urbanism, health, culture, entertainment, recreation, and tourism. One of the main GCC ambitious targets is to move away from oil dependence and redirect the oil and gas exploration, currently used for domestic energy consumption, to other higher-value uses. The GCC states' ambitions are based on their respective visions and investment plans. Figure 1 summarizes the GCC national visions and plans. The KSA based its Vision 2030 (KSA-2030, 2017; KSA-NTP, 2017) on its status as the heart of the Arab and Islamic worlds, its determination to become a global investment powerhouse, and transforming its strategic location into a global hub – an epicenter of trade and a gateway to the world.

UAE Vision 2021 (UAE-2021) focuses on world-class healthcare, competitive knowledge economy, safe public, and fair judiciary: a cohesive society and preserved identity, a first-rate education system, and sustainable environment and infrastructure. The State of Qatar Vision 2030 (QATAR-2030) embraces economic development, shifting from an oil-based economy to a knowledge-based economy, social development, human development, and environment development. The State of Kuwait Vision 2035 (KUWAIT-2035) is set to move the country toward Smart Kuwait with major sustainable environment projects and targets effective public administration, sustainable diverse economy, enhanced infrastructure, sustainable living environment, high-quality healthcare, creative human capital, and enhanced global position. The Sultanate of Oman Vision 2020 (OMAN-2020) is also interested in a sustainable and diversified economy with the objective of creating a stable and holistic economy, encouraging private sectors, reducing dependency on oil, enhancing the standard of living while reducing disparities between regions, and upgrading of Omani workforce and knowhow skills. Sultanate of Oman is preparing its future vision 'Oman 2040' adopting three main priorities, namely 'People & Society,' 'Economy & Development,' and 'Governance and Institutional Performance.'

Many countries have already been engaged in SCC (Wafula, 2016) with a focus on key primary indicators like smart governance, smart people, smart economy, smart environment, smart mobility, and smart living (Cohen, 2018; Giffinger, 2014). Two models have been used for smart cities frameworks, namely the Greenfield model where cities are created from scratches such as Masdar City in the United Arab Emirates or King Abdullah Economic City (KAEC) in KSA, and the Brownfield model where cities are built based on existing cities such as Masqat in Oman. According to the literature, many Brownfield model-based smart cities and Greenfield model-based smart cities are expected to be built by 2025 in the GCC region with Saudi Arabia and the UAE investing approximately 49.3 billion USD (Gulf, 2017). Currently, the GCC countries are adopting international best practices and leveraging opportunities to transform into digital government models, with themes like smart cities, smart tourism, next-generation care, open government data (OGD), classroom of the future, smart government, and future of mobility (McKinsey, 2016; Willen, Nasr, Zuazua., Lohmeyer, & Romkey, 2018). For instance, in their recent series of transformations, GCC governments are pouring investment into smart cities in accordance with the recent aggressive government-driven goals (Deloitte, 2017). The state of Qatar and Dubai are both on the fast track preparing to host the 2022 FIFA World Cup and the 2020 World Expo. Urban planners in most GCC cities are already including smart city infrastructure and features in their new transformation plans and visions. Investment in a smart city is positioned to grow in most of the GCC with an increased focus in KSA and UAE. The Smart Dubai initiative seeks to make the city as the smartest city in the world (Wafula, 2016).

In recent years, cities in the GCC states have significantly grown, especially with the launching of large cultural, sports and entertainment projects, and many public parks and gardens – all set to provide citizens with the best services, good quality of life, healthy lifestyle, and attractive living environment, equipped with good recreations areas where people would practice sport, enjoy life with families and friends, and promote social well-being (UAE-2021). For example, Qatar is constructing a smart city 'Lusail' (Lusail, 2018) with sustainable facilities, leisure spots, residential buildings, commercial towers, public marinas, and smart disposing of waste, capable of hosting 250,000 residents. Kuwait has also planned the building of nine smart cities and is currently developing its first smart and environment-friendly city 'Saad Al-Abdullah' (Bayat, 2018; ICS, 2017). Oman is also focusing on smart key initiatives and has, so far, developed a smart city platform as a knowledge-sharing consortium to drive its smart city initiatives (Oman, 2018). McKinsey Global Institute study 2018 ranked Abu Dhabi first among cities in Africa and the Middle East as the most developed technology bases while Dubai was ranked first in the best cities deploying the greatest number of applications in all domains (McKinsey, 2018). Masdar City is powered by clean energy and integrates tradition with modern architectural techniques capturing winds and offering naturally cooler outdoor public spaces (http://masdar.ae).

However, Saudi Arabia has launched a 500 billion USD mega-city project known as 'NEOM' (Neom, 2017) running on 100% renewable energy, with the first phase expected to be completed in 2025. KAEC includes King Abdullah Port and is designed on social, economic, and environmental sustainability principles. The Saudi 'Smart City' initiative is set to gradually target five Saudi cities by 2020 and to apply smart city concepts as one of a series of municipal transformation projects with the objective of bringing together various 'smart' components: buildings, transport systems, security and safety services, communication systems, drainage networks, street lighting systems, and an emergency response system. Five smart initiatives will be adapted to apply the concept of smart cities, including smart parking, smart lighting systems, smart solid waste disposal, smart cameras, and environmental pollution monitoring tools.

GCC and Smart Village Initiatives

ICT alone cannot be the solution for smart villages as attention to native histories, traditions, spiritual values, and human capital preparation is necessary. This means that research on smart cities, smart megacities, or smart villages rather require social and political sensitivity as well as forward-thinking (Visvizi & Lytras, 2018b). It is with this idea in mind and with attention to the location of the GCC countries in an arid region characterized by severe drought climate and water supply shortages and reliance on desalination to provide water to their population that one ought to approach and evaluate the shift to smart villages. Some GCC countries use treated wastewater for irrigation. For example, Kuwait and Saudi Arabia reuse about 50% of their total wastewater while Bahrain and Qatar reuse about 10–15% (Efron, Fromm, Nataraj, & Sova, 2018). Sustainable agriculture growth and rural development are considered as key to alleviating poverty in rural areas. In the GCC countries, agriculture contributes only 10–15% of the GCC GDP (NCB, 2010). The scarcity of fresh water and energy is becoming more pronounced because of the growth of population and economic prosperity. However, sustainable agriculture is needed to ensure food security, ensuring access to affordable, reliable, and modern energy. Interestingly, the GCC countries moved from food self-sufficiency to food security policies (Efron et al., 2018; NCB, 2010) with their improved healthcare during the last decade. Moreover, in the field of education, the GCC countries have shown an increased number of private universities along with the international alliances with a positive impact on their educational landscape. Digital divide, however, remains a concern in rural areas especially in the domain of ICT and high-speed internet.

GCC countries are also making progress toward sustainable development. For example, in cooperation with the United Nations Development Program (UNDP), the Saudi National Space Strategy (2030) was updated to support the implementation of the Vision 2030 in the provinces, towns, and villages of the Kingdom integrating both the SDGs and the new urban agenda objectives (KSA-UN, 2018). Some villages already exist; however, more are related to tourism as is the case with 'Rijal Almaa Village,' 'Ghat Heritage Village,' or 'Habala Villages,' given that they contain historical, cultural, and natural elements. Other heritage tourist villages exit in the rest of the GCC countries. In Abu Dhabi, 'Hydra Village' is the first community to be built on the smart city concept. It is an eco-friendly, self-sustained, and small-scale city. In the case of Saudi Arabia, more work is currently being undertaken on smart networking of villages. In collaboration with Electricity de France Company (EDF) and under a joint venture program (SVRG, 2018), on March 2017, the Saudi Electricity Company (SEC, 2017) signed a cooperation agreement to establish a smart network village research laboratory (grid concept) in the style of a similar research laboratory in France.

In view of the above, one needs to note that smartization of cities and villages alongside the translation of smart concepts in GCC urban and rural contexts require serious attention to the indigenous makeup of culture, religion, history, and overall environment. This implies revisiting the approaches being used so far toward the development of rural areas. This entails to the discussion of community preparedness to move along the Government SMART agendas, and certainly calls for improving communications channels, education, and training. For example, a field study of 17 major cities whose population makes up nearly 72% of the country's total population on the preparedness level of Saudi cities to transform into smart cities reveals that there is currently disparity in the cities' preparedness to shift to smart cities, with Makkah coming in first, followed by Riyadh, Jeddah, and Madinah and Ahsa (Arab News, 2017). And while there is a continuous interest, the GCC region still lags behind other developed economies as reported by Emmanuel Durou, Partner and Leader for TMT in Deloitte Middle East (Deloitte, 2017). This is understandable given the many challenges and problems affecting the conceptualization and development of smart villages.

One of the challenges is the research gap between implementation of smart city and smart village. More research is needed to further understand the peculiarities of each and the relevant appropriate strategies. According to Lytras and Visvizi (2018), research focusing on transposing insights from smart cities research to the context of rural areas (i.e., smart village research) is just at the beginning. Another problem relates to the need to bridge the various distant communities, connect their endeavor and interest, and remove the distances isolating people, which necessitates effective use of digitization using adapted programs, training, and concepts, models of entrepreneurship and innovation, and business solutions, and overall improvement of rural communities. Speaking of government, Belcaid (Nagraj, 2017) highlights another problem pertaining to the awareness of smart cities, which also applies to the smart villages. While he acknowledges that Dubai is in the lead and regularly communicates about its smart city plans (Abbas, 2018), he noted that "other cities in the region may not be as advanced. Or worse, becoming smart is not yet on their agenda" (Nagraj, 2017).

Moving toward smart villages, however, several problems surface, including, for instance, data sharing among government agencies and with the public, lack of skillful workforce in the areas critical to digital transformation, and community connections. As Salem noted (Nagraj, 2017), smart transformation needs an understanding of data science and data-related practices, which appears to be deficient in the current workforce for him. Salem states, "You have major islands of information that are detached from each other and that makes it almost impossible to create efficiencies and develop any smart initiative." The needs still persist on reviewing the conditions of rural areas, building of smart villages and empowering rural communities, and providing basic life services and more efficient environmental management, the guidance of national and district officials, as well as the private sector stakeholders in the design, development, and establishment of green smart villages. Communities in rural areas need to be gradually educated and trained in green and smart building and living – for instance, including green, yet smart concepts of houses building and related infrastructure, consumable assets (water, biogas, etc.), incorporation of ICT, sustainable agricultural practices, and efficient water and energy management, sanitation, and hygiene. Waste management is also necessary as a solution to farming, recycling of substances into productive resources, and preventing the use of toxic chemicals in sewage treatment plants, not to forget about effective community settlement and economic production.

GCC countries have the capacity to attract a greater number of global citizens, foreign investors, and global corporations. However, this requires attracting more global corporations and foreign direct investment in addition to increasing the global impact of local corporate groups and localizing investments and gaining the support of the private sector. Investment should be made toward innovation, enabling supply and driving innovation demand. There is a necessity to build human capital networks and expand cultural offerings and improve cities' ability to retain global talent. They need to strengthen the connecting infrastructure to support urbanization, including roads and transmission grids, educational institutions, and diverse value networks. Having said that, efficient physical infrastructure is necessary to support any realization of smart village and sustainable development in general. Priority should be given to electricity and water networks and services, road network, transportation system, and wastewater treatment and sewage facilities with attention to the construction of biogas production units in villages and remote communities. Similarly, government needs to further invest in renewable sources of energy and water – including water recycling and desalination using renewable sources of energy – as positive measures toward sustainable patterns of production and consumption, leading to enhancing efficiency in the use of water, energy, and food, and a shift toward renewable water and energy sources and integrated waste management techniques and practices.

Conclusions

First one needs to carefully approach the comparison of GCC performances on smart villages in such a way that political and sociocultural variances are not excluded; actually, one should undertake such a task, with countries of close or similar sociohistorical composition, as well as economic and development indexes, while at the same time recognizing the increasing investments made toward improvement of citizens welfare, infrastructure, and levels of sustainable developments. In spite of those continuous investments being put forward, however, more is needed toward revitalizing rural services through digital and social innovation, integration of ICT in the design, and delivery of various services including smart networks, digital facilities, and e-governance, health, social services, education, energy, transport, retail, business networking, and overall life improvement. GCC countries need to invest in developing relevant sound policies and tools necessary for sustainable smart villages while providing continuous support to the current initiatives being taken so far, as well as the possible initiatives showing interest into this direction. Given their common cultural and sociohistorical background and their homogenous visions of development, the countries of the GCC would benefit from a shared holistic plan for smart villages for sharing and exchanging creative innovations and experience. More research is needed, especially with regard to the problems and challenges affecting the development transitions, and more importantly perhaps, the adaptation of smart villages and ways to design concepts adaptable to the local GCC contexts of change and circumstances of development − not to forget about the need for promoting holistic approach with effective policies and strategies toward smart villages.

Acknowledgment

The authors gratefully acknowledge the support of the College of Engineering and the General Education at Effat University, Jeddah, Kingdom of Saudi Arabia.

References

Abbas, W. (2018, August 13). Dubai marches towards becoming the smartest city. *Khaleej Times*.

Abdulla, M. (2015). Gulf Cooperation Council initiatives: Towards a coherent integrated plan for utilizing renewable solar energy. In D. Bryde, Y. Mouzughi, & T. Rasheed (Eds.), *Sustainable development challenges in the Arab States of the Gulf* (pp. 193−211). Berlin: Gerlach Press. doi:10.2307/j.ctt1df4hx6.14.

Albino, V., Berardi, U., & Dangelico, R. M. (2015). Smart cities: Definitions, dimensions, performance, and initiatives. *Journal of Urban Technology*, *22*(1), 3−21. doi:10.1080/10630732.2014.942092

Al-Ebrahim, A. (2017). Integrating new and renewable energy in the GCC region. *Renewable Energy Integration*, 189−197. doi:10.1016/b978-0-12-809592-8.00014-7

AlMunajjed, M., & Sabbagh, K. (2011). *Youth in GCC countries, meeting the challenge*. Report of Ideation Center Insight, Booz & Company.

Ankur, S., Shyamu, G., & Gaurav, T. (2018, March). Smart villages: Progress of Indian era − Today's need. *International Journal for Research in Applied Science & Engineering Technology*, 6(3), 783−788.

Arab News. (2017). *Smart city initiative launched*. Retrieved from http://www.arabnews.com/node/1087402/saudi-arabia. Accessed on April 8, 2019.

Bayat, E. (2018). *Planned city projects in Kuwait you need to know about*. Retrieved from https://ehsanbayat.com/4-planned-city-projects-in-kuwait-you-need-to-know-about/. Accessed on November 29, 2018.

Bled. (2018). *Bled Declaration*. Retrieved from http://pametne-vasi.info/wp-content/uploads/2018/04/Bled-declaration-for-a-Smarter-Future-of-the-Rural-Areas-in-EU.pdf. Accessed on November 10, 2018.

CGIAR. (2017). *Research program on climate change, Agriculture and Food Security*. Retrieved from https://www.cgiar.org/wp/wp-content/uploads/2018/09/CCAFS_AnnualReport_2017.pdf. Accessed on December 2018.

Cohen, B. (2011). *The top 10 smart cities on the planet*. Retrieved from http://www.fastcoexist.com/1679127/the-top-10-smart-cities-on-the-planet. Accessed on February 12, 2014.

Deloitte. (2017). *National transformation in the Middle East: A digital journey*. Deloitte & Touche (M.E.) Report.

Efron, S., Fromm, B. G., Nataraj, S., & Sova, C. (2018). *Food security in the Gulf Cooperation Council*. Report as part of the RAND Corporation external publication series.

European Commission. (2017). *EU action for smart villages*. Retrieved from https://enrd.ec.europa.eu/news-events/news/eu-action-smart-villages_en. Accessed on November 15, 2018.

European Court of Auditors. (2018). *Renewable energy for sustainable rural development: Significant potential synergies, but mostly unrealised*. Special Report No. 5, European Union.

European Union. (2016). *Cork 2.0 Declaration: A better life in rural areas*. Retrieved from https://enrd.ec.europa.eu/sites/enrd/files/cork-declaration_en.pdf. Accessed on December 10, 2018.

Gevelt, T. V., Holzeis, C. C., Fennell, S., Heap, B., Holmes, J., Depret, M. H., & Safdar, M. (2018). Achieving universal energy access and rural development through smart villages. *Energy for Sustainable Development*, 43, 139−142. doi:10.1016/j.esd.2018.01.005

Giffinger., R. (2014). *Smart cities ranking of European medium-sized cities*. Centre of Regional Science, Vienna UT. Retrieved from http://www.smart-cities.eu/download/smart_cities_final_report.pdf

Gulf Business. (2017). *Smart cities: Is the GCC seeing the transformation?* Retrieved from https://gulfbusiness.com/smart-cities-gcc-seeing-transformation. Accessed on November 14, 2018.

Gulf Business NCB Capital. (2010). *GCC agriculture, economic research*. Retrieved from https://www.gulfbase.com/ScheduleReports/GCC_Agriculture_Sector_March2010.pdf

Hogan, P., Crețu, C., & Bulc, V. (2016). *EU action for smart villages*. Brussels: European Commission.

ICRISAT. (2016). *Building climate-smart villages: Climate and crop modeling approach*. Retrieved from http://www.icrisat.org/wp-content/uploads/2016/11/Building-Climate-Smart-Villages.pdf. Accessed on November 14, 2018.

ICS. (2017). *Innovative Cities Summit – Kuwait*. Retrieved from https://isocarp.org/events/innovative-cities-summit-kuwait-2017/. Accessed on November 30, 2018.

IEEE. (2018, November 11). *Smart village*. Retrieved from http://ieee-smart-village.org/

ITU. (2014). *Smart sustainable cities: An analysis of definitions*. ITU-T Focus Group on Smart Sustainable Cities. Retrieved from http://www.itu.int/en/ITU-T/focusgroups/ssc/Pages/default.aspx. Accessed on November 11, 2018.

KSA-2030. (2017). *Vision 2030*. Retrieved from http://vision2030.gov.sa/en. Accessed on May 2018.

KSA-NTP. (2017). *National Transformation Program*. Retrieved from http://vision2030.gov.sa/sites/default/files/NTP_En.pdf. Accessed on October 15, 2018.

KSA-UN report. (2018). *Towards Saudi Arabia's Sustainable Tomorrow*. First Voluntary National Review 2018–1439, UN High-level Political Forum. Retrieved from https://sustainabledevelopment.un.org/content/documents/20230SDGs_English_Report972018_FINAL.pdf. Accessed on September 15, 2018.

Kuwait. (2018). *Kuwait National Development Plan*. Retrieved from http://www.new-kuwait.gov.kw/home.aspx. Accessed on September 15, 2018.

Lara, A. P., Costa, E. M., Furlani, T. Z., & Yigitcanlar, T. (2016). Smartness that matters: Towards a comprehensive and human-centered characterization of smart cities. *Journal of Open Innovation: Technology, Market, and Complexity*, 2(1), 1–13. doi:10.1186/s40852-016-0034-z

Lusail. (2018). *Smart City Vision*. Retrieved from http://www.lusail.com/the-project/smart-city/. Accessed on December 10, 2018.

Lytras, M. D., & Visvizi, A. (2018). Who uses smart city services and what to make of it: Toward interdisciplinary smart cities research. *Sustainability*, *2018*(10), 1998. doi:10.3390/su10061998

Mckinsey & Company. (2018). *Smart cities: Digital solutions for a more livable future*. Mckinsey Global Institute. Retrieved from https://www.mckinsey.com/industries/capital-projects-and-infrastructure/our-insights/smart-cities-digital-solutions-for-a-more-livable-future. Accessed on November 30, 2018.

McKinsey, Digital McKinsey. (2016). *Digital Middle East: Transforming the region into a leading digital economy*. Retrieved from goo.gl/H2rQBhcontent_copy. Accessed on Nonember 28–29, 2018.

Mora, L., Bolici, R., & Deakin, M. (2017). The first two decades of smart-city research: A bibliometric analysis. *Journal of Urban Technology*, 24(1), 3–27. doi:10.1080/10630732.2017.1285123

Nagraj, A. (2017). *Smart cities: Is the GCC seeing the transformation?* Gulf Business. Retrieved from https://gulfbusiness.com/smart-cities-gcc-seeing-transformation/. Accessed on December 10, 2018.

Neom. (2017). *Saudi Arabia plans to build futuristic city for innovators*. Retrieved from https://phys.org/news/2017-10-saudi-arabia-futuristic-city.html. Accessed on October 24, 2017.

Oman. (2018). (2018, December 10). *The Official Oman eGovernment Services Portal*. Retrieved from http://www.oman.om/wps/portal

OMAN-2020. (2018, October 2). *Vision 2020*. Oman. Retrieved from https://www.scp.gov.om/en/Page.aspx?I=14. Accessed on November 25, 2018.

QATAR-2030. (2018, October 2). *Vision 2030*. Retrieved from www.qatar.vcu.edu/files/Appendix_M_2030_Qatar_National_Vision.pdf. Accessed on November 25, 2018.

Ramachandra, T. V., Hegde, G., Subhash Chandran, M. D., Kumar, T. A., & Swamiji, V. (2015). Smart village: Self-sufficient and self reliant village with the empowerment of manpower (rural youth) through locally available natural resources and appropriate rural technologies. Smart Village, Energy & Wetlands Research Group, CES, IISc, ENVIS, Technical Report: 90, Energy & Wetlands Research Group.

Ramady, M. A. (2012). *The GCC economies: Stepping up to future challenges*. New York, NY: Springer.

Saudi Press Agency (SPA). (2017, April 20). Smart city initiatives. *Arab News*.

SEC, Saudi Electricity Company. (2017). *SEC and EDF sign two agreements*. Retrieved from https://www.se.com.sa/en-us/pages/newsdetails.aspx?NId=567. Accessed on October 29, 2018.

SVRG. (2018, December 2). *Smart villages research group, smart villages, new thinking for off-grid communities worldwide*. Retrieved from http://e4sv.org/about-us/what-are-smart-villages/. Accessed on December, 10, 2018.

The Economist. (2010). The GCC in 2020: Resources for the future. A report from the Economist Intelligence Unit Sponsored by the Qatar Financial Centre Authority. Retrieved from http://graphics.eiu.com/upload/eb/GCC_in_2020_Resources_WEB.pdf. Accessed on November 29, 2018.

UAE-2021. (2018, October 2). *Vision 2021*. Retrieved from https://www.vision2021.ae/en. Accessed on November 25, 2018.

UN. (2010). *United Nations UNCTAD: Renewable energy technologies for rural development*. United Nations Conference on Trade and Development, UNCTAD Current Studies on Science, Technology and Innovation, No 1.

UN. (2015). *Transforming our world: The 2030 Agenda For Sustainable Development*. Retrieved from https://www.un.org/pga/wp-content/uploads/sites/3/2015/08/120815_outcome-document-of-Summit-for-adoption-of-the-post-2015-development-agenda.pdf. Accessed on December 10, 2018.

UN. (2017). *UN-Habitat, implementing the new urban agenda by strengthening urban-rural linkages – Leave no one and no space behind*. United Nations Human Settlements Program (UN-Habitat).

UN. (2018). *The United Nation world urbanization prospects 2018, urban and rural population*. Retrieved from https://population.un.org/wup/. Accessed on December 10, 2018.

UN-SDG. (2018). *The Sustainable Development Goals Report 2018*. Retrieved from https://unstats.un.org/sdgs/files/report/2018/TheSustainableDevelopmentGoalsReport2018-EN.pdf

Visvizi, A., & Lytras, M. D. (2018a). It's not a fad: Smart cities and smart villages research in European and global contexts. *Sustainability*, *2018*(10), 2727, doi:10.3390/su10082727

Visvizi, A., & Lytras, M. (2018b). Rescaling and refocusing smart cities research: From mega cities to smart villages. *Journal of Science and Technology Policy Management*, *9*(2), 134–145.

Wafula, M. (2016). *ICT policies and plans for transition to smart and sustainable development in the Arab region*. Report, International Telecommunication Union (ITU) and Telecommunication Development Bureau (BDT).

Webb Henderson. (2015, November 20). *How do smart cities succeed?*. Legal and Regulatory Advisors. Retrieved from https://webbhenderson.com/wp-content/uploads/2016/01/Smart-Cities-June-2015.pdf

Willen, B., Nasr, A., Zuazua, M., Lohmeyer, R., & Romkey, M. (2018, November 20). *Global cities of the future – A GCC perspective*. World Government Summit, in collaboration with ATKearney. Retrieved November 2018 from https://www.worldgovernmentsummit.org/api/publications/document?id=9e6c7dc4-e97c-6578-b2f8-ff0000a7ddb6

Zavratnik, V., Kos, A., & Duh, E. S. (2018). Smart villages: Comprehensive review of initiatives and practices. Retrieved from https://doi.org/10.20944/preprints201807.0115.v1

Chapter 12

Smart Villages: Mapping the Emerging Field and Setting the Course of Action

Miltiadis D. Lytras, Anna Visvizi and György Mudri

Set against the backdrop of a process that swipes across the world, that is, rapid urbanization coupled with depopulation of rural areas, the objective of this volume was to ignite a discussion on the future of villages, and especially on the wellbeing of its citizens. Captured by the powerful concept of smart villages (Visvizi & Lytras, 2018a), the imperative behind this volume was to engage the leading scholars, think-tankers, policy advocates, civil servants and, indeed, policy-makers in a discussion centered around the question of what we can do together to navigate the daunting implications of depopulation of villages across the European Union (EU) and beyond. Throughout the book, the implications of progressing depopulation of rural areas have been discussed. At the same time a variety of viable recommendations and feasible policy actions meant to alleviate the situation have been suggested. Seen in this way, this volume has a very strong normative and prescriptive value. On the one hand, by proposing a comprehensive approach to the concept of smart village, it highlights that we cannot afford only an academic, concept-based, debate on smart villages. Rather, this comprehensive approach this volume advocates makes it mandatory that the stakeholders listen to each other and seek ways of addressing problems and nascent risks. Accordingly, the approach to smart villages that this volume embodies advocates that the identification of real needs and problems faced by inhabitants of villages is followed by the construction of sound conceptual frameworks and by evidence-driven targeted policymaking. Action has to be taken now. Even if consensus around this plea consolidates at the EU level, more effort is needed to translate it in local-level policymaking and the mechanisms that underpin it.

This edited volume addresses several big questions, including that of the appropriate set of policies and strategies that the EU member-states can devise to literally induce life in villages across the EU. It looks at the intersection of diverse policy fields and competencies of the EU and its member-states to map the options and alternatives at the stakeholders' disposal (see also Zavratnik,

Kos, & Stojmenova Duh, 2018). In this context the role of the Common Agricultural Policy (CAP) has been stressed. The potential inherent in instruments and mechanisms specific to Cohesion Policy has been flagged up. The chapters included in this volume look at very specific domains that define smart villages' functioning and operation, including local government services, energy sustainability, precision farming, entrepreneurship development, as well as questions of employment, investment in rural areas, environmental protection, and community building.

To what extent and how advances in information and communication technology (ICT) can be employed to address these issues and related challenges? What are the obstacles and enablers of implementing smart solutions that from a technical and technological point of view are at the reach of our hand (Lytras & Visvizi, 2018)? Is there a hope for villages in the EU? Will the next generations be given the opportunity to develop bright memories of pastoral country side? Will they have the opportunity to serve as guardians of the heritage our villages entail?

Advances in ICT, including big data, data analytics, data mining, sensors, virtual reality, augmented reality, 5G technologies, block chain technology etc. redefine the landscape of our daily life (Sicilia & Visvizi, 2018; Visvizi & Lytras, 2019). They have profound impact on how we live, work, spend our free time, and travel. They influence the way we receive our medical treatment, the degree of safety (not only online) we experience, the way we communicate, etc. (Chui et al., 2017; Lytras & Papadopoulou, 2017; Visvizi, 2015). Arguably, this snapshot overview of ICT's impact on our lives is equally relevant in villages (Visvizi & Lytras, 2019). The chapters included in this volume alluded, frequently implicitly, how ICT-enhanced solutions might benefit villages and their inhabitants across the EU. For instance, Daniel Azevedo, in Chapter 6, makes a very compelling case for precision farming. More work needs to be done to explore at length in which ways ICT and the existing conceptual framework and policy instruments could be employed to exploit the potential ICT has for villages and its inhabitants. This is the forthcoming step of our research agenda.

The important message that this volume makes is that the concept and approach to rural development and policymaking, captured by the term smart village, go beyond the techno-hype. The unique value proposition inherent in the concept of smart village is that it seeks to do much more than to showcase how sophisticated ICT can be employed in a village context. In fact, a very strong normative component rests at the heart of smart village research. It is the conviction that while ICT-enhanced approaches are necessary to ensure well-being, safety, and happiness of the inhabitants of villages, it is our obligation – in the spirit of, say it, solidarity – to ensure that villages are revitalized and regain the luster of attraction they deserve.

Accordingly, the chapters comprising this volume make a case for a multifaceted, ICT-conscious, comprehensive, and integrated strategy aimed at revitalizing rural areas across the EU. Comparative insights also from outside the EU, including the Gulf Cooperation Council (GCC) countries and Korea place the challenge of rural areas' decline and effective use of ICTs for the sake of their revival in a unique analytical perspective. As a result, this volume serves as a primer for all those willing to understand the idea behind the concept of smart village, its relevance and most profoundly the promise of undertaking targeted and focused action right now.

By engaging in interdisciplinary, concept-driven and case study-led conversations among experts, academics, and practitioners, this volume addresses several questions pertinent to the future of villages across the EU. By so doing, this volume provides intrinsic knowledge of the field, understanding of concepts and their applicability, an overview of tools and strategies already implemented, and a thorough insight into synergies that emerge as the interest in smart villages matures. As a result, this volume is bound to serve not only as the first source of reference in the smart villages debate, but also a one-stop shop for all those who seek to make sense of the connection that unfolds among problems and challenges villages are exposed to today and the promise ICT may generate in this regard.

Having said that, more research, dialogue and policy analysis is necessary if the message that this book entails is to be communicated effectively to all stakeholders. As mentioned above, the concept of smart villages, very consciously so (Visvizi & Lytras, 2018b), avoids the ICT-hype. At the heart of the smart village concept and approach lies the wellbeing of inhabitants of villages, young and old, tired and motivated, disappointed and still hopeful. Rather than showcasing what sophisticated technology can do, this volume elaborated on what needs to be done and what is feasible. Aware of the promise and potential ICT-bears, the next step will be to match the needs and challenges to policy actions and ICT-enhanced solutions. The Editors of the volume and so the contributing authors have already engaged in this discussion, hopeful to be able to frame it and consolidate it through a joint research project.

References

Chui, K. T., Alhalabi, W., Pang, S. S. H., Pablos, P. O. D., Liu, R. W., & Zhao, M. (2017). Disease diagnosis in smart healthcare: Innovation, technologies and applications. *Sustainability*, *9*(12), 2309.

Lytras, M. D., & Papadopoulou, P. (Eds.). (2017). *Applying big data analytics in bioinformatics and medicine*. Hershey, PA: IGI Global. doi:10.4018/978-1-5225-2607-0

Lytras, M. D., & Visvizi, A. (2018). Who uses smart city services and what to make of it: Toward interdisciplinary smart cities research. *Sustainability*, *10*(6), 1998, doi:10.3390/su10061998

Sicilia, M. A., & Visvizi, A. (2018). Blockchain and OECD data repositories: Opportunities and policymaking implications. *Library Hi Tech*, doi:10.1108/LHT-12-2017-0276.

Visvizi, A. (2015). Safety, risk, governance and the Eurozone crisis: Rethinking the conceptual merits of 'global safety governance'. In P. Kłosińska-Dąbrowska (Ed.), *Essays on global safety governance: Challenges and solutions* (pp. 21–39). Warsaw: Centre for Europe, University of Warsaw, ASPRA-JR, 2015. ISBN: 83-89547-24-4.

Visvizi, A., & Lytras, M. D. (2018a). It's not a fad: Smart cities and smart villages research in European and global contexts. *Sustainability*, *10*(8), 2727, doi:10.3390/su10082727

Visvizi, A., & Lytras, M. D. (2018b). Rescaling and refocusing smart cities research: From mega cities to smart villages. *Journal of Science and Technology Policy Management*, *9*(2), 134–145, doi:10.1108/JSTPM-02-2018-0020

Visvizi, A., & Lytras, M. D. (Eds.). (2019). *Politics and technology in the post-truth era*. Bingley: Emerald Publishing.

Zavratnik, V., Kos, A., & Stojmenova Duh, E. (2018). Smart villages: Comprehensive review of initiatives and practices. *Sustainability*, *10*(7), 2559.

Index

Adarsh Gram Yojana programme (Government of India), 106
AgriChemWay, 21
Agriculture 2.0, 85
Agriculture 3.0, 85
Agriculture 4.0, 85
Agriculture and food production
 challenges with, 84–85
 climate change, 84–85
 EU agri-food sector, 84
 EU code of conduct on agricultural data sharing, 92
 precision agriculture (PA)/ farming (PF), 85–90
 bottlenecks, 91–92
 examples, 87–88
 forestry, 88
 importance to smart villages, 90–92
 precision livestock farming (PLF), 88–90
 rate of technology uptake by farming community, 84, 85, 88–90
 future trends, 90
 technological transformation and, 93

Bioeconomy, 24
Bioenergy Villages project, 103
Bled Declaration, 2018, 2–3, 15, 52, 128, 158
Broadband infrastructure, 77

Clean energy driven smart villages, 105–106
Clean energy technologies, development of
 IT skills and, 104–105
 rural high-speed internet links and, 104
Cohesion policy, 2–3, 19–20
Committee of Regions (CoR), 13
Common Agricultural Policy (CAP), 2–3, 14, 16–19, 94, 141, 173–174
Community-Led Local Development (CLLD), 128
Community-led Local Development (CLLD) approach, 64
Connectivity, 30
Consultative Group for International Agricultural Research (CGIAR) Research Program on Climate Change, 158
Cork 2.0 Declaration 2016, 2–3, 8, 14, 15, 21, 25, 51–52, 128
COWOCAT project, 76

Depopulating villages, 1–2, 67
 Fouskari, xv
 Greek villages in Peloponnese, 3
Digital economy, 76–77
Digitalisation of smart villages, 104–105

Digitalization of cities, 14–15
Digital Single Market, 16, 94–95

Eigg Heritage Trust project, 118–119
Embeddedness, 30
ESPON 2020 Cooperation Programme, 128
'EU Action for Smart Villages', 15–16, 63
EU agricultural and food production sectors, 84
EU General Data Protection Regulation (GDPR), 92
EU policies for rural areas, improving coherence of, 21–22
EU policies for smart villages
   challenges for implementation at local level, 40–43
     access to institutions, 42
     technical infrastructure, 42
     use of resources, 42–43
   cohesion policy, 19–20
   Common Agricultural Policy (CAP), 17–19
   context and definition, 37–40
   research and innovation, 20–21
     Digital Innovation Hubs, 20
     LIAISON, 20
     LIVREUR, 20
     ROBUST, 20
     RUBIZMO, 20
     RURACTION, 20
     SIMRA, 20
European Commission, 2–3
   on barriers to adopt digital technologies, 94–95
   Basic Payment Scheme (BPS) applications, 91
   on broadband coverage in Europe, 91
   General Data Protection Regulation (GDPR), 92
   long-term climate vision for 2050, 85
   Mobile Agricultural Robot Swarms (MARS) experiment, 90
   new Skills Agenda for Europe, 95–96
   'The Internet of Food and Farm 2020' project, 89
European Economic and Social Committee (EESC), 14
European Innovation Partnership (EIP-AGRI), 18, 21, 24–25, 72–73, 94
European Network for Rural Development (ENRD), 13–14, 15, 21, 24–25, 118
European Union, 1
   Clean Energy Package, 105–106
   development of rural high-speed internet links, 104
   differences between the rural and urban territories, 50
   funding for non-agricultural activities, 13
   population living in rural and urban areas, 49–50
Eurostat, 49–50
EU strategy of technological transformation, 95

Gandhi, Mahatma, 158
Ghat Heritage Village, 164
Green Revolution, 85
Gulf Cooperation Council (GCC) countries, 155–156, 174–175

digital government models, 162–163
National Transformation Plans, 161–162
population distribution and estimation (2018-2030), 160, 161
smart village initiatives, 164–166
transformational context, 159–164

Habala Villages, 164
Horizon Europe, 24–25
Howard, Ebenezer, 55
Hungarian 'grandma application,', 120
Hydra Village, 164

'I can do this for you' philosophy, 9
IEEE Smart Village, 128
Information and communication technology (ICT), 3–4, 7, 49, 50, 55, 85, 149–150, 165–166, 174–175
  advances in, 174
Innovations, 31, 41
  small- and middle-sized enterprises (SMEs), role of, 114
International Crops Research Institute for the Semi-Arid Tropics (ICRISAT), 158

Korea, 174–175
  agricultural and rural investment plans in, 144
  agricultural policy in, 139–140, 143
  agricultural sector in, 139
  block grant system, 145–148
  Comprehensive Rural Village Development Program (CRVDP), 145
  rural policy, 140–141, 142–143
  rural spaces in, 139
  rural tourism, 144, 145
  rural tourism policy in, 148
  rural tourism programs in, 146–147
  6th industrialization, 149
  smart farming, 149–150
  Smart Village Project in, 143–148

LEADER approach to smartness, 63–65, 68, 128
  budget, 73
  common features shared, 68
  innovation, 72–73
  integrated development, 70–71
  partnership and community empowerment, 71
  place-based approach, 69–70
  distinctive characteristics between smart villages and LEADER, 69
  geographical scope, 73
  local actions groups (LAGs), role of, 71, 72–73, 78, 79
  modern rural development policy paradigm and, 65–68
  opportunities for sustainable development, 68
  potential role of LEADER, 73–77
  as animator and facilitator of community processes, 74–75

as enabler for transition, 76–77
as laboratory of innovation for a transitional change, 75
role in digitization, 76–77
supporting scaling-up of initiatives through cooperation, 75–76
regulatory barriers in, 78–79
LEADER Local Action Groups, 24

Masqat, 162–163
Mobile Agricultural Robot Swarms (MARS), 90

Nanotechnology, 90
National Strategy for Inner Areas, Italy, 67
NEOM project, 163–164

Paris Agreement, 100, 101
Plan for Rural Digitisation, Finland, 67
Power technologies for smart villages, 102–104
biogas, 102–103
biomass, 103–104
wind and solar thermal, 104

Regional Policy, 2–3
Relatedness, 30
Resilience, 50–51
Revitalization, 10
Rezo Pouce project, 76
Rijal Almaa Village, 164
Rural development, 31
concepts
endogenous approach, 33–35
evolution of policies, 36–37
mixed exogenous/endogenous approach, 35
neo-endogenous approach, 35
policy, 37
EU, 37–38
role of agriculture, 83–85
Rural proofing, 21–22, 25

San-Car, 120
Saudi National Space Strategy (2030), 164
Saudi National Transformation Program 2020 and Vision 2030, 159
Saudi 'Smart City' initiative, 163–164
Service hubs in rural Flanders, 119
Slovenia
Blatna Brezovica, 131
concept of smart village, 128–129
Distributed Hotel Konjice, 130
FabLab Network, 132–134
Hotel of Good Teran, 131
innovation, 132–133
landscape of, 129
Ljubljana, 129
mobility services, 131–132
projects Agrotur I and Agrotur II, 131
rural areas
challenges of, 127
problems of, 125–127
SaMBA, 131–132
smart rural development in, 125, 133
tourism, 129–131
village Padna, 130–131
Small and medium-sized enterprises (SME), 9
Small- and middle-sized enterprises (SMEs), 111

ability to develop smart solutions, 119–120
components of intelligent or smart solutions, 116–117
connection between smart villages and, 111
effectiveness of, 112
project examples, 118–119
role in European economies, 112–114
role in innovation, 114
role in sustainability, 114–115
rural, 112
smart solutions for, 115
sources of financing for, 117–118
Smart cities, 156–159
Smart Dubai initiative, 162–163
Smart Eco Social Villages, 15
Smart Fab Village, 133–134
'Smart' food production processes, 90–92
Smartness of a settlement, 14
Smart rural development, 31
implementation of, 31
notion of innovativeness, 31–32
Smart Specialisation under Cohesion policy, 16
Smart strategies for smart villages, 52–53
methodology for design of, 53–59
actors involved in process, 54–55
coherent structure of government and territory, 58
commitment to social cohesion and development, 58
connections with other rural areas and cities, 58–59
environmental sensitivity and responsibility, 57–58
first step in defining, 56–59
garden city model, 55, 56
theoretical framework, 55–56
work and innovation, 58
Smart Village Model, 51–52
Smart villages, 1
advantages of, 56
benefits of living in, 57
challenges and corresponding actions, 2
definition of, 38–39, 111
EU policies for, 14–16
European Commission definition, 3
European policy discussions, 14
five avenues of
communication, 24
innovation, 24–25
integration, 23
rural proofing, 25
simplification, 23–24
implementation under integrated rural development framework, 67
importance of, 4–5
as a policy objective, 13–14
rationale behind, 1–2
relevance, 2–4
three pillars, 3–4
Smart Villages Portal, 15, 118
Solar power, 8–9, 99–100
Solar-powered smart villages, 100–102
digital nature of solar power, 102, 104–105
rationale for, 101

roof panels for, 101–102
State of Kuwait Vision 2035 (KUWAIT-2035), 162
State of Qatar Vision 2030 (QATAR-2030), 162, 163
Sultanate of Oman Vision 2020 (OMAN-2020), 162, 163
Sustainable Development Goals (SDGs), 65, 100, 157, 164
Sustainable Smart cities (SSC), 155, 162–163

Thematic group (TG) on Smart Villages, 38, 40
Thematic Working Group, 15
"Three magnets" theory, 55, 56
Toolkit for the development of Smart Green Villages, Rwanda, 106
"The 25 Most Innovative AgTech Startups in 2018,", 120

UAE Vision 2021 (UAE-2021), 162

Venhorst Declaration, 2017, 22
Village angels, 115–116

'WAB' initiative, 77